Flight Testing Homebuilt Aircraft

Flight Testing
Homebuilt

Aircraft

VAUGHAN ASKUE

Iowa State University Press / Ames

Vaughan Askue is an aeronautical engineer and manager of customer development at Sikorsky Aircraft in Stratford, Connecticut.

The purpose of this book is to provide information on the basic methods and techniques of flight testing homebuilt aircraft. The user of this information assumes all risk and liability arising from such use. Neither Iowa State University Press nor the author can take responsibility for the actual operation of a homebuilt aircraft or the safety of its occupants.

First edition, 1992

Library of Congress Cataloging-in-Publication Data

Askue, Vaughan
    Flight testing homebuilt aircraft / Vaughan Askue.—1st ed.
      p.   cm.
    Includes index.
    ISBN 0-8138-1308-5
    1. Airplanes, Home-built—Flight testing.   I. Title.
TL671.7. A85 1992
629.134'53—dc20   91-26885

[CONTENTS]

They shall mount up

with wings as eagles —ISAIAH 40:31

The human race has always dreamed of flying. The way was pioneered by test pilots. Whether good or bad, professional or amateur, lucky or unlucky, they built a library of knowledge and technique that we can use today to fly better and more safely. I have simply tried to communicate a small part of that library.

The idea for *Flight Testing Homebuilt Aircraft* occurred on a summer evening in August 1983. A number of us were sitting around the hangar, after a Sailplane Homebuilders Association workshop, talking airplanes. At the workshop an FAA inspector had discussed gaining approval for homebuilt aircraft. His talk was spiced with clear examples of how not to do it. Our group in the hangar continued the discussion, each adding horror stories to the group's collective wisdom. As I listened, it struck me that most of these incidents could have been prevented through the use of flight test techniques that were generally accepted within the industry, but apparently not common knowledge to homebuilders, and that the best way to present these techniques to the average homebuilder would be in a book. I broached the idea to Jim Loyd and Dave Hudnut (then Eastern Region vice president of SHA) and their response was enthusiastic. Driving home full of excitement, I was ready to get the ball rolling.

The ball would roll for eight years. Over those eight years many people contributed to the book. In some cases it was knowingly, through encouragement, advice, or knowledge. In other cases people contributed unknowingly as their experiences, through word of mouth or articles, helped me discover pitfalls as well as solutions to problems. For example, Rudy Opitz showed a mutual friend how to use the wing bending natural frequency to detect structural damage.

Ralph Lee and Jim Loyd helped me get started with encouragement and advice. My wife, Sandy, tolerated long evenings while I was handwriting the first draft. When the draft was typed, Tova Clayman invested many hours editing to ensure that my engineering lingo was intelligible.

Once the book had found a publisher a new cast of characters helped me through the final stretch. Mary Curran completed the final smoothing of the text and helped me fix remaining organizational errors. Reviewer William Kershner discovered a flaw in the chapter on spins. Jim Patton, retired NASA test pilot, gave willingly his time and expertise to make it right.

To all of these I give my thanks. Without their help, *Flight Testing Homebuilt Aircraft*, even if completed, would have been a lesser product.

[INTRODUCTION]

"*Test pilot*"

The name conjures up images of macho excitement equalled by few other professions. The words inspire images of Tom Cruise, the "right stuff," leather jackets, and screaming dives in which only the test pilot's skill, strength, and incredible courage stand between a successful pullout and a large smoking hole in the ground. The test pilot role has ranked in Hollywood with that of the secret agent for mystique. Unfortunately, test flying in Hollywood fiction is just that—fiction.

Today's test pilot is a highly trained professional, probably middle-aged, a college graduate, and a family person. The best test pilots approach their profession with the intellect and intensity of an orchestra conductor. This is not a bad analogy, because a test pilot for a major test program is supported by tens, sometimes hundreds, of experts ranging from mechanics to engineers and by masses of special equipment from telemetry systems to spectral analyzers.

There are other test pilots, however, who don't fit this picture. Every year hundreds of homebuilt aircraft are completed, signed off by the FAA, and flown. The pilots of these airplanes may not have the training, experience, or extensive support of the military and industrial test pilots, but they are just as much test pilots as the professionals.

Unfortunately, lack of knowledge and training means that few homebuilt aircraft are properly tested. Simply flying off the hours of the FAA restricted period accomplishes little more than burning fuel. Furthermore, if approached with little knowledge and a poor attitude, it can actually be quite risky. The results can include accidents that wreck airplanes and hurt people or airplanes that fly poorly because the builders do not know how to correct their problems.

All is not lost, however, because amateur builders can use a few basic methods and techniques to do a perfectly adequate job of flight testing their own airplanes. The purpose of this book is to give you, the builder, the tools to test your creation safely with professional results.

Although we will talk a lot about specific techniques and tools, the most important thing to understand is the basic definition of flight testing. Flight testing is the step-by-step process of learning an aircraft's limitations, defining and fixing its problems, and determining its capability and optimum flying techniques. Each step in the test process is designed to prepare both the pilot and the airplane to progress to the next step with a minimum of risk.

Understanding and working with this philosophy is important because there are few hard-and-fast rules in flight testing. Every airplane, every pilot, every test situation is different and will require different treatment. This book provides guidelines derived from years of painful experience of the professionals, but it is not a cookbook. It cannot cover every situation you may run into, nor will it always provide the solution to your specific problem. Use this book as a guide to develop your own test program. The book should help you decide what you are trying to accomplish and the best way to accomplish it.

I will examine the flight test program by breaking it into four phases: preflight, first flight, envelope expansion, and performance. Preflight is the phase of flight testing that prepares the aircraft to move under its own power. It includes such things as rigging, weight and balance, and fuel-flow checks. The first-flight phase is intended to build up to and include the first flight. It includes taxi tests, land-backs, and the first flight itself. Envelope expansion is the process of determining the aircraft's envelope and limitations and of making the aircraft fly properly within that envelope. The last phase, performance testing, is used to determine what the airplane can do and the optimal way to get the most from it.

This sounds like an awful lot of work. Why bother? Is it worth it? The restricted period mandated by the FAA for homebuilt airplanes is forty hours or less. It certainly would be easier to bore holes in the sky for forty hours. But what a waste! If you can discipline yourself to flight-test your airplane properly, you will increase your knowledge of your airplane and its capability tremendously. As a bonus, you will also notice an enormous improvement in your own piloting skills. You may even find this business so fascinating that you spend much more time testing and developing your airplane than required. Are you still with me? Let's get started.

Sonerei II L / *all metal*

Christen Eagle / *steel tube and fabric*

Boredom Fighter / *wood and fabric*

Flight Testing Homebuilt Aircraft

Falco / *all wood*

[CHAPTER 1]

How to Begin

The flight test process should begin when you first spread out your newly acquired plans on the kitchen table and begin figuring out how to build your airplane. You'll want to ensure that a number of design features are present, both for safety and to make the most commonly required modifications easier. If you are building from a kit or well-established plans, you may not feel you have the freedom to make modifications. However, most of the modifications we will discuss will not alter the basic design of your airplane. They may even improve features that are often poorly specified by the plans, if they are called out at all. Besides, this is your airplane, and most homebuilders are inveterate improvers anyway, especially when they can improve safety or accessibility.

In order to make our discussion more logical, let's discuss the design features system by system.

Cockpit

The first consideration in a cockpit is space. You will need enough space to be comfortable for fairly long periods of time, to reach and operate all the necessary controls, and to escape from the airplane if necessary. Most builders discover cockpit space problems when construction is well under way. This leads either to a lot of cutting and splicing and rebuilding (and swearing) or to living with discomfort or awkwardness

3

for the life of the airplane. An easier way is to discover these problems before serious construction is under way and to fix them as you build.

If you are building an airplane from plans or a kit and you can find someone near you who is building the same airplane, go and sit in it, even if it isn't flyable. If you can't find one to sit in, or if you are designing a unique airplane, a cockpit mock-up is a good idea. It doesn't need to be fancy, but it should have the shape of the instrument panel, the important controls, and the door or opening you'll use to get in or out. It should also be dimensionally accurate and strong enough for you to get in and out and move around inside without breaking anything.

Whichever approach you are taking, first put on your parachute and helmet and climb in. You should fit comfortably. This is not meant to be funny. If you are a particularly odd size (or any part of you is), or if you are building one of those homebuilts with an oddly shaped cockpit, this could be a major problem. If you seem to fit, sit there for a while and hangar-fly. All the important controls should be easy to reach with no stretching, twisting, or bending over. Controls that require you to lean forward and look down are particularly dangerous because rotating the head forward while the aircraft is maneuvering can upset your middle ear and cause vertigo. You should be able to get full travel on all the controls without your body or the airplane getting in the way.

After sitting in the cockpit for a while, you should still be comfortable. You should not feel stiff or sore, and there should be no hard parts of the airplane poking into you. If you feel as if you need to stretch and can't, you probably have a limb in an awkward position.

You should also have enough room so you can look back over your shoulder to the 5- and 7-o'clock positions reasonably comfortably without hitting your helmet on the structure. There should be at least an inch of clearance between your helmet and the canopy.

Now try to get out of the cockpit with your helmet and parachute on. The canopy latch should be easy to reach and natural to operate. You should be able to climb out easily without wiggling or snagging on parts of the airplane.

This may seem like a lot of work even before you start to build. The time is well spent, however, because cockpit problems have a strong effect on how safe your airplane will be and on how enjoyable it will be to fly. It is far more efficient to find a problem and design a fix from the beginning than to try and solve the same problem when your ship is half-built.

There are several other items that will want to provide in your cockpit. Every homebuilt must be equipped with a four- or five-point harness. The crotch strap in a five-point harness is intended to keep you from "submarining" or sliding out from under the lap strap under high loads. While always desirable, the crotch strap is particularly important with the semireclined seating found in sailplanes and some racers. A shoulder harness is cheap insurance, so get a good, new one with no visible wear. Never use a harness that has been involved in an accident. An aircraft harness is designed to stretch under load. Once stretched, its strength is reduced and it must be scrapped. The harness should be firmly attached to the primary structure, as shown in Figure 1.1. The lap belt should approximately bisect the angle between the seat back and the seat pan, and when you are seated it should rest across your hip bones.

The conventional wisdom among homebuilders has been that the shoulder harness should be mounted level with or above the pilot's shoulder. This is because in the case of a violent stop the pilot's upper body pitches forward over the lap strap and is restrained only by the shoulder harness. If the shoulder harness is mounted too far below the level of the pilot's shoulder, it can apply damaging, compressive loads to the pilot's back as the upper body swings forward. Many homebuilders have intentionally mounted their shoulder harnesses level with or above the level of the pilot's shoulders to avoid this problem. This is good practice. Research in the automotive industry (Studer 1989) has shown

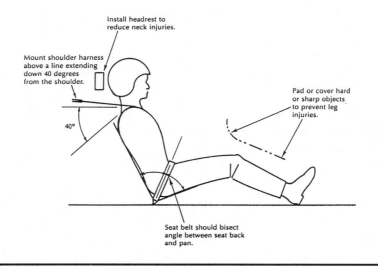

Fig. 1.1. Making a safe cockpit

that a downward angle of up to 40 degrees is acceptable, as shown in Figure 1.1.

You will not always be flying with a parachute, and other people, who are different sizes, will be flying your airplane so it is good practice to provide for seat adjustment. One good way of doing this is to provide seat and back cushions of varying thicknesses. These can be held in place with strips of Velcro. One of these cushions can replace your parachute as a back support when the test flying is finished.

The design of cockpits and instrument panels is a complex science and deserves a separate book, but there are several things you should be sure you have: airspeed indicator, altimeter, sideslip indicator (ball-race), accelerometer (g-meter), tachometer, compass, oil pressure and, temperature gages, fuel gage, and ammeter or voltmeter. Turn-and-slip (in place of the ball-race), rate-of-climb, and heading indicators are nice to have but not all that important. A fuel pressure gage will be required if a fuel pump is installed, and a manifold pressure gage will be necessary if you have a constant speed prop. The instruments mentioned above that are not required by the Federal Aviation Regulations (FARs) are intended to help you monitor onboard systems or to measure things you can't see such as load factor. You will be flying by visual flight rules (VFR), so extensive flight instrumentation is really not necessary and adds weight and cost. A gyro horizon, for example, is heavy, expensive, and unnecessary in most homebuilts.

Determining the layout of instruments and switches in your cockpit is another time when a mock-up can pay off. A mock-up panel will, first of all, give you a feel for how good your visibility is going to be. All the instruments and switches you need to use in flight should be easy to see and reach without stretching or craning.

The instruments you use every few seconds in flight, such as the airspeed or the sideslip indicators, and those that will be critical for test points, such as load factor, should be right in front of you and as high as possible without blocking your visibility. Instruments used less often, such as the altitude and rate-of-climb indicators, can be lower on the panel and farther from the pilot's centerline.

By the same token, switches used often and those critical to flight should be close to the pilot's throttle hand, and switches used once or twice per flight can be located almost anywhere that they can be reached without too much stretching or twisting. Switches critical to flight are for items such as the fuel boost pump, the landing light, and the master switch. These should be located together, close to the throttle, and they should go *on* in the same direction.

There is an accepted convention for the direction of switch motion in an airplane. In order to turn a switch on, thus activating a device or function, the pilot's hand must move forward or up. For example, if a toggle switch is located on a center console with the toggle facing upward, the toggle should move forward to *on*. If the same switch is located on an overhead console, facing down, the toggle should still move forward to *on*.

Engine

The biggest enemy of your engine is heat, and the only ways you have of measuring excessive heat are oil temperature and cylinder-head temperature. Both should be measured, but cylinder-head temperature is the more accurate and responds more rapidly. Because each homebuilt engine installation is unique and there is no sure way to tell ahead of time which cylinder will be the hottest, you should measure them all. A separate gage for each cylinder is ideal but expensive. A thermocouple on each cylinder, wired through a selector switch to a single indicator (engine analyzer), is the next best option. If this is impossible, at least put thermocouples on the two cylinders farthest away from the cooling air inlet. Thermocouples should also be installed on the hottest plug on a given cylinder. This is normally the plug facing away from the cooling flow. For an upflow cooling system, this would be the upper plug, and for a downdraft system, the lower plug.

Engine thermocouples normally are bimetallic washers that replace the sparkplug washers on the cylinders you wish to measure. When the bimetallic washer is heated, a voltage is created that is measured by the head-temperature indicator. This voltage is small so the measurement is strongly affected by small changes in resistance in the wiring, switches, or meter. For this reason all the components of the head-temperature measuring system must be installed as a matched set. NEVER cut, splice, or alter the wiring to a thermocouple. If the wires are too short, new ones of the right resistance must be obtained from the manufacturer. If you are installing several thermocouples through a switch to the same meter, check with the manufacturer to ensure that the parts you are going to use together are compatible.

Control System

The two things you should make sure are provided for in the control system are adequate stops and a means for rigging. The control system should be equipped with adjustable stops in the cockpit for the stick and pedals and at the control surfaces for the elevators and rudders. Some different types of stops are shown in Figure 1.2.

The control system should be equipped with sufficient adjustment to allow proper initial rigging and subsequent rigging changes if tests show they are necessary. If you have a cable control system, the drawings should show a turnbuckle on each cable running to each control surface. Thus, there should be two turnbuckles in the elevator-control system and two in the rudder system. The aileron system should have a minimum of three, and four are better. The biggest thing to remember is that you will

Pedal

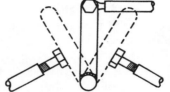

Torque tube

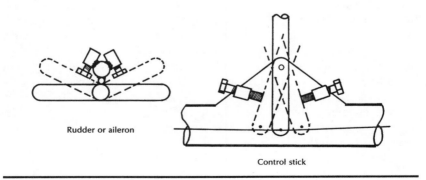

Rudder or aileron

Control stick

Fig. 1.2. Control stops. Note that ALL control stops must be safetied.

have to use these turnbuckles to adjust the control position and the cable tension with the aircraft fully assembled. All turnbuckles must be accessible when each control is at its extreme as well as at its mid position.

Pushrod systems are normally rigged using adjustable rod-end bearings. To ensure proper geometry of the system, put an adjustable bearing at each end of each rod. Adjustable rod-end bearings are normally equipped with a safety gage to prevent you from extending them too far and reducing their strength. This safety gage usually consists of a small hole drilled in the barrel of the bearing. If a piece of safety wire can be slipped through the hole, the bearing is extended too far; either the adjustment must be made elsewhere or the pushrod must be removed and rebuilt to a longer length.

While examining the control system, keep in mind that all the fittings, bellcranks, rod ends, and pulleys will need to be inspected periodically and sometimes adjusted. All must be accessible. As many points of adjustment as possible should be located within the fuselage where they can be reached easily. If turnbuckles must be located within the wings, place them at the horn or bellcrank so one inspection panel allows access to both.

Electrical and Fuel Systems

There are two primary concerns pertaining to the electrical and fuel systems. The first is the separation of the components of each system from each other and from anything that would damage them. The second is the ability to cut off the fuel and electrical supplies quickly and easily.

The fuel system must be equipped with a cutoff valve that you can easily reach with your throttle hand. You should not have to switch hands or reach across the cockpit to shut off the fuel supply in an emergency. All fuel lines should be isolated from the airframe to prevent them from being worn by vibrating against the structure. The best way of securing fuel lines is with rubber-lined hose clamps bolted to the primary structure. It may be necessary to use standoffs between the clamps and the structure to ensure adequate separation between the line and the nearest piece of airplane. Grommets or phenolic blocks must be used wherever lines pass through a bulkhead.

Like the fuel valve, the electrical master switch must be located where you can switch it off quickly and easily with your throttle hand. It should operate a supply relay located as close to the battery as possible. This relay should cut off all power to the aircraft except, of course, to the master switch and the magneto *p* leads. The location of the relay close to the battery is very important because it is your primary means of dealing with a short anywhere in the electrical system. The battery stores enough energy to start a healthy fire, and if a fire occurs between the battery and the relay you will have no means of cutting off the flow of energy.

Although it is not a part of the electrical system, you should always install a means of breaking the circuit (either a fuse or a switch) between an impulse electric tachometer and the magneto. If the tach shorts internally it will act as a *p* lead and ground out the magneto. In a single-mag engine this will immediately lead to an awful silence. You can then restore power by pulling the fuse or switching off the tach.

As with the fuel system, all electrical lines should be secured with clamps or tie-wraps and kept away from other objects. Wires should be protected by grommets or phenolic blocks anywhere they pass through a bulkhead.

As I mentioned at the beginning of this section, it is good practice to keep fuel and electrical lines as far from each other as possible. One way of doing this is to pass the fuel lines down the sides of the airplane and confine the wiring to the fuselage centerline, or vice versa.

Weight Control

While building your airplane, you will probably be tempted to add all sorts of goodies. You will see pictures of Oshkosh grand champions boasting umpteen layers of hand-rubbed enamel and interiors that would make a Rolls Royce proud. You will see visions of full gyro panels, radio stacks, LORAN-C receivers, electric starters, and auxiliary fuel tanks. The temptation is strong—resist it. Weight is the enemy of performance, and the smaller the airplane the more it suffers from a pound of excessive weight. The result of poor weight control is often a two-place airplane that will carry only one and even then performs poorly.

Consciously decide what you want your airplane to be before you start, and don't expect it to be all things. If you want a grand champion or an airplane that will fly from New York to Los Angeles nonstop, expect it to be expensive and time-consuming to build and to have a poor payload. If you want an airplane that performs well so you can take your friends for rides, keep the weight down by keeping the airplane simple.

For the same reason, don't mess with structure. Most aircraft structures look deceptively fragile to the layperson. Adding another layer of glass, an uncalled-for gusset, or the next-larger wall thickness of tubing will only hurt performance. If the design you are building has a good history, trust the drawings. You will come out with a lighter, more efficient airplane.

Reference

Studer, Richard. 1989. "Safety Restraints," letter to the editor, *Sport Aviation* (February), 61.

Further Reading

Andrews, David. 1989. "Safety Device," letter to the editor, *Soaring* (March), 3.
Bingelis, Tony. 1984. "The Effects of Engine Compartment Heat," *Sport Aviation* (December), 12.
_____. 1985. "Homebuilt Aircraft Interiors, Part 1," *Sport Aviation* (January), 49.
_____. 1987a. "Minimizing Your Battery Problems," *Sport Aviation* (January), 35.
_____. 1987b. "An Introduction to Kits," *Sport Aviation* (March), 33.
_____. 1987c. "Your Brake Installation," *Sport Aviation* (October), 31.
_____. 1988. "A Look at Some Wiring Practices," *Sport Aviation* (July), 27.
Bingham, Neil. 1986a. "Engine Installation in a Sport Plane," *Sport Aviation* (March), 54.
_____. 1986b. "Light Is Better," *Sport Aviation* (December), 39.
Blackburn, Albert. 1986. "Canard Canard," *Aerospace America* (August), 38.
Burns, B. R. A. 1985. "Canards: Design with Care," *Flight International* (February) 23, 19.
Chandler, Richard, and Mike Mahoney. 1985. "Restraint System Basics," *Sport Aviation* (January), 35.
Collinge, George. 1984. "Is a Horizontal Tail Necessary?" *Sport Aviation* (April), 27.
Crawley, Ed, and R. John Hansman. 1989. "Improving the Crashworthiness of Your Sailplane," *Soaring* (June), 19.

Experimental Aviation Association. *Custom Built Sport Aircraft Handbook.*

"Flight Testing with Models." 1965. *Air Progress* (January), 26.

Hall, Stan. 1987a. "Dynamic Modeling," *Sport Aviation* (July), 31.

_____. 1987b. "Testing of Structurally Scaled, Sacrificial Models as an Aid to Full Scale Design," *Sport Aviation* (August), 59.

Larsen, Chuck. 1985. "Simplified Installation Criteria for Shoulder Harnesses," *Sport Aviation* (January), 60.

Mohler, Stanley. 1987. "Benefits of Crashworthiness Confirmed at the Indianapolis Speedway," *Sport Aviation* (April), 62.

_____. 1988. "Crashworthiness from the Human Standpoint," *Sport Aviation* (November), 34.

Owen, Ben. 1987a. "Installing a Canopy Safety Latch," *Sport Aviation* (March), 39.

_____. 1987b. "Aircraft Instrument Markings," *Sport Aviation* (August), 46.

_____. 1987c. "Color Them Cool," *Sport Aviation* (September), 35.

Powell, Lyle, Jr. 1987. "Fuel Systems for Homebuilt Airplanes," *Sport Aviation* (April), 29.

Robertson, David. 1988. "Rx for a Small Instrument Panel," *Sport Aviation* (February), 59.

Schiff, Barry. 1983. "Beyond the Blueprints," *AOPA Pilot* (October), 43.

Taylor, Molt. 1984. "Overweight," *Sport Aviation* (February), 38.

_____. 1987. "Models for Test and Designing Homebuilt Aircraft," *Sport Aviation* (January), 58.

Zeisloft, Harry. 1984. "Avoiding Fuel Contamination," *Sport Aviation* (November) 58.

[CHAPTER 2]

System Tests

As your airplane approaches completion, you must perform a series of system tests and checks to ensure that all its major systems are functioning properly. You should complete the checks and log the results before the FAA inspector comes for the final inspection. Don't skimp on these checks; they are vital to the safe flight of your airplane. These tests cover fuel flow, fuel system calibration, airspeed system leaks, control system rigging and interference, weight and balance, and landing gear rigging.

Fuel Flow

The fuel-flow check demonstrates that your fuel system will deliver enough fuel flow for your engine to develop full power in the most critical flight conditions. For gravity-fuel systems in conventional tractor airplanes, the most critical condition is a nose-high pitch attitude, which would occur during a stall recovery or an aborted landing. The nose-high attitude of the aircraft reduces the height of the tank with respect to the carburetor and thus decreases the fuel pressure at the carburetor inlet.

If your airplane is equipped with an unconventional fuel system, the most critical condition may not be a wings-level, nose-high attitude. For example, with a pusher engine the fuel tank is normally located forward of the engine, so the worst case would be a steep glide when the nose is down and the fuel tank is at its lowest with respect to the carburetor.

Before you start, remember that this test involves uncontained fuel. It is thus fairly dangerous and requires some special precautions. It must be done outside, in low winds, and away from cars, structures, or any flammable material. It requires two people, and a good fire extinguisher should be handy. Smoking, of course, should not be allowed anywhere nearby.

The plain wings on most homebuilts stall at an angle of attack of about 15 degrees. You also need to allow several degrees of nose-up attitude in case the airplane is climbing at or near a stall angle. You should therefore position the aircraft so that the wing chord is 17 to 20 degrees nose-up. For a tailwheel airplane, this will mean putting the main gear up on blocks, or the tailwheel in a hole, or both.

You will require 1½ to 2 gallons of fuel, a bucket, a stopwatch, and a stool to hold the bucket just below the carburetor. Calibrate the bucket with a measuring cup and mark it at the 1-gallon level (this is 16 cups or 128 fluid ounces, if you are using your kitchen measuring cup). The bucket must be kept level so that the calibration and the test read the same.

Add enough fuel to the tank so that there is some standing fuel in the tank. Then disconnect the fuel line at the carburetor and allow fuel to run out until the flow stops. The amount of fuel remaining in the system at this point is referred to as the unusable fuel, and it should also be in the system when the aircraft is weighed. Make sure the bucket is empty. Then cap the line with your thumb and add about 1½ gallons of fuel to the tank. Place the bucket under the fuel line, release your thumb, and start the stopwatch. Stop the watch when the fuel reaches the 1-gallon mark in the bucket. Repeat the test several times and average the results. The following equation will tell you what the resulting fuel flow is.

$$\text{Fuel flow (gallons/hour)} = \frac{\# \text{ gallons} \times 60}{\# \text{ minutes}}$$

The fuel flow should equal or exceed the fuel flow called out in the engine operator's manual or the engine specs at full power. This test works all right for fuel capacities of 10 gallons or more. For smaller fuel systems, use about 10 percent of the fuel capacity instead of 1 gallon and modify the equation accordingly.

If your fuel system fails the test, first check for restrictions in the fuel passages. Look for kinks in the lines, undersized fittings, a blocked fuel vent, a fuel valve not fully open, or dirt or debris in the lines, fittings, or tank strainer. If you have none of these problems, you can improve fuel flow by installing larger lines and fittings, thus reducing line losses. You can also raise the tank with respect to the carburetor to increase the pressure (although this is tough to do) or install a fuel pump. The fuel pump option is normally a last resort because of reliability problems.

Note that all homebuilts except those with fuel pumps need to have their tank vents facing into the slipstream. This is required because most carburetor float bowls are pressurized by ram air passing through a vent inside the carburetor inlet. Thus, pressure within the bowl will increase with speed. If this is not balanced by ram air in the fuel tank, the carburetor will run lean at high speeds.

Fuel System Calibration

A fuel system calibration is required because most fuel measurement systems are not linear and some vary with attitude. It is important to know the amount of fuel onboard for weight, c.g. (center of gravity), and safety purposes.

It is easiest to do the fuel system calibration at the airport after the airplane is assembled and ready to be run. Airport fuel pumps are calibrated accurately by each state's Board of Weights and Measures, so you can use the fuel pump itself as a calibration standard.

A tailwheel airplane presents a special problem for fuel system calibration. The large difference between the ground and in-flight pitch attitudes usually affects the fuel gage reading, so that it gives different answers in flight from those it gives on the ground. To make sure you always know what the gage is telling you, you will need to do two calibrations, one with the tailwheel on the ground and one with the tailwheel in the normal flight position. The easiest way to do this is to find or make up a stand just tall enough so that the airplane is in a normal flight position when its tailwheel is on the stand. Putting the stand under the tailwheel after each tail-down calibration point is taken allows you to measure the indicated fuel at both attitudes.

To begin, place the airplane in the normal ground position and add enough fuel so that there is standing fuel in the tank. Disconnect the fuel line so that fuel runs out. After the flow has stopped, reconnect the fuel line. The fuel that is now in the system is the unusable fuel and is considered part of the aircraft's empty weight.

If your aircraft has an electric fuel-indicating system, it is easiest to calibrate to the marks already given on the fuel gage (¼, ½, etc.). Put the tail up on the stand so that the fuselage is in the flight position. Fill to the first increment on the gage (quarter mark) and note the number of gallons. Remove the stand, put the tailwheel on the ground, and again note the gage position. Repeat the process at each indicated gage position up to the full mark. Then plot the results as shown in Figure 2.1. Using this calibration, you can now accurately determine the fuel load both on the ground and in flight.

If you are using a sight gage or another indicator to which you can apply your own scale, you can build the calibration right into the graduations you put on the scale. In the case of a taildragger, which may show a significant difference in readings for ground and in-flight attitudes, you may be able to use two scales, one for each attitude. In this case you should calibrate with fixed increments of fuel and mark the gage accordingly. These increments should not be larger than ¼ of the tank's capacity. If, for example, a 2-gallon increment is chosen, put 2 gallons in the tank with the airplane in the tail-down position and mark the amount of fuel on the scale. Put the tail in the flight position and mark the fuel level on the other scale. Repeat at even increments.

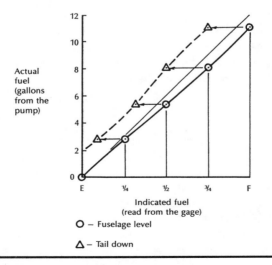

Fig. 2.1. Calibration of a premarked fuel gage

Airspeed System Leaks

Internal leaks are a major source of airspeed system errors and may be the reason that some airplanes are "slow." To detect and eliminate these errors, you must perform a system leak check before you fly your aircraft. Obtain about a foot of tubing that will fit tightly over your pitot tube. Fold one end over twice and seal it with a couple of wood blocks and a C-clamp, as shown in Figure 2.2.

With someone watching the airspeed indicator, carefully slip the open end of the tubing over the pitot tube until the indicator reads about the expected cruise speed. If the tubing doesn't slide on easily, lubricate the tube with a little soapy water. Monitor the airspeed

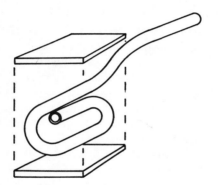

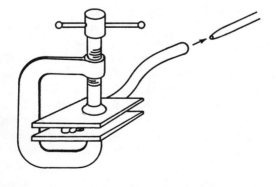

Fig. 2.2. Pitot system leak check

indicator over the next ten minutes, occasionally tapping it to prevent instrument friction from showing a deceptively small drop.

The airspeed should not decrease more than 2 knots over ten minutes. If it does, you probably have a leaking joint or joints in the pitot plumbing. Tighten all the fittings and try again. If a system is particularly troublesome, check that you have a tight fit between the pitot tube and the tubing and a good seal on the end of the tubing. Also check that the pipe thread fitting in the indicator is not leaking. If the leak is large, some soapy water brushed on the joints may show the leak; however, the pressures we are dealing with are so small this is not always reliable.

Control System Rigging

Control system rigging consists of adjusting the control stops, turnbuckles, and rod-end bearings and making any other adjustments so as to allow proper control travel. Although control angles are normally specified on the drawings, a full procedure is rarely specified, so here is the procedure you should use for the control systems shown in Figure 2.3. As an aside, I think it's a lot easier to rig and do interference checks on a fabric-covered airplane before it is covered.

Before beginning rigging, you need to have a method of measuring control deflections. Professionals normally do this with a device called a blade protractor, an expensive precision instrument with an accuracy of 0.2 degrees, which is far more precise than we need. The simplest alternative is to make a set of rigging fixtures or templates. These are plates of aluminum or thin plywood designed to fit snugly on the wing or fixed surface. They are equipped with a scale positioned so that the trailing edge of the movable surface indicates the deflection. They are easy to build and, if made carefully, will be perfectly adequate for your purposes. Typical examples are shown in Figure 2.4.

To rig the ailerons, block the stick in the center position by inserting wood or aluminum blocks in the roll-stick stops and by clamping down with the stop adjustments. Tighten the two upper aileron turnbuckles until the ailerons are at the neutral position called out by the plans. The ailerons are then usually, but not always, in alignment with each wing's trailing edge. Tighten the lower aileron turnbuckle(s) until proper cable tension has been achieved. Recheck to make sure that tightening the

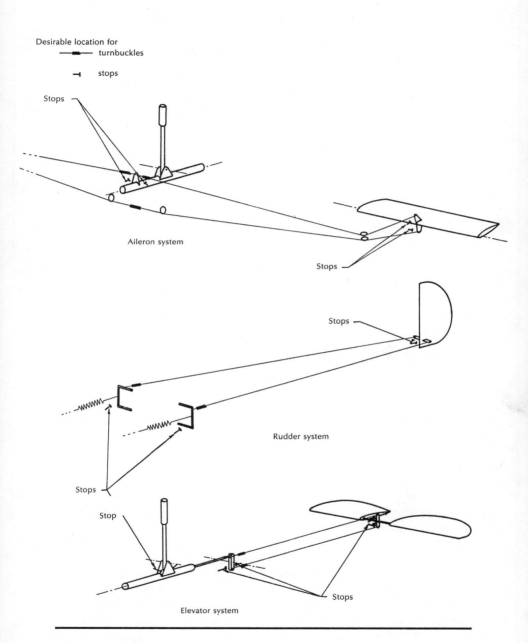

Desirable location for
━━■━━ turnbuckles
⊣ stops

Stops

Aileron system

Stops

Stops

Rudder system

Stops

Stop

Stops

Elevator system

Fig. 2.3. Control systems

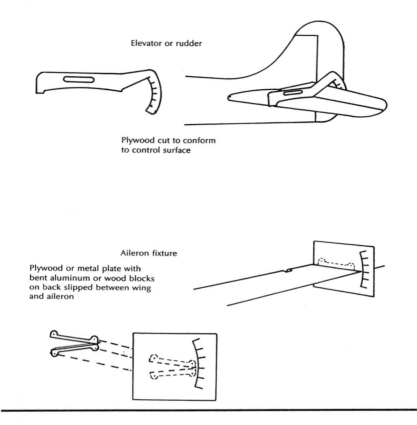

Elevator or rudder

Plywood cut to conform
to control surface

Aileron fixture

Plywood or metal plate with
bent aluminum or wood blocks
on back slipped between wing
and aileron

Fig. 2.4. Control rigging fixtures

lower turnbuckles hasn't pulled both ailerons down below the specified mid position. If it has, adjust both upper and lower turnbuckles in small increments to bring the ailerons to neutral position while retaining cable tension. Now release the blocks and adjust the stick stops to set the maximum up- or down-aileron angle called out by the drawings.

The elevators and rudder are normally set up a little differently. To check the rigging, open the cockpit stops for the longitudinal stick and the pedals so that the controls are limited only by the stops at the control surfaces. Adjust these stops so that the control travels are those called out by the drawings. Adjust the turnbuckles so that the tension is proper and the stick and pedal travels are comfortable. Using a feeler gage, adjust the cockpit stops so that with moderate pressure on the control the stops show 0.010 inch clearance.

INTERFERENCE CHECK

After rigging, it is important that you do an interference check. Have someone sit in the cockpit and cycle the controls to their extremes while you watch each linkage and control mechanism for any possibility of interference with other systems on the airframe. No control system component should come within 0.10 inch of the airframe structure at any point or within 0.50 inch of any system component. The person moving the controls should also check to ensure that the cockpit controls do not interfere with the instrument panel, the seats, or any other cockpit hardware. When checking the clearance between the stick or throttle grips and the surrounding structure, be sure to wear heavy gloves so you don't get any nasty surprises when you start winter flying.

Weight and Balance

The business of determining a homebuilt's empty weight and c.g. has, I'm sure, always seemed like another exercise in "dumb government require-ments." Nothing could be further from the truth. As you will find later, an airplane's gross weight and c.g. have a profound effect on perform-ance, stability, control, and ground handling. In fact, altering the c.g. may be one of your most effective tools in making your airplane fly well. The first step in calculating the gross weight and c.g. for any configuration is to determine an accurate weight and c.g. for the empty airplane.

Empty weight and c.g. are normally calculated by measuring the aircraft's weight at two points along the length of the airframe and then using some basic moment calculations to determine how much the airplane weighs and at what point it balances. Normally the points used are the contact points for the main landing gear and the nosewheel or tailwheel. If you remember your high school physics, you'll know you need two pieces of information to calculate a moment—the force and a moment arm.

Each moment arm is calculated from a reference point normally located out in front of the nose of the aircraft. This simplifies the math by ensuring that the arms of the measured loads and the resultant c.g.

location are all positive numbers. The reference point's location in space is specified in terms of some fixed location on the aircraft, such as the firewall or the leading edge of the wing, as shown in Figure 2.5. All points on the aircraft, including the c.g., are then located in terms of their "station"—the distance in inches from the reference point. It is normal practice to assign a nice round station number to the fixed fuselage location to keep the math simple, as shown in Figure 2.5. For example, if the firewall is defined as station 100, the reference point would be a point in space 100 inches forward of the firewall. A point 23.7 inches aft of the firewall would be at station 123.7.

When selecting the fixed location for station zero, do not use a point on the engine or propeller, such as the prop flange. In case you discover a c.g. problem serious enough to require moving the engine to rebalance the airplane, your reference point would move also, creating a major opportunity for errors.

The locations of the landing-gear contact points can sometimes be taken from the drawings if they are accurate enough and if you haven't altered your airplane. It is good practice, however, to actually measure the contact points for your airplane. You can do this most easily by leveling the aircraft over a hard floor and measuring the distance between the fixed point (the firewall) and a chalk line drawn between the main landing-gear contact points. For example, let's assume that the line between the main gear points measures 20 inches aft of the firewall and the tailwheel measures 237.5 inches aft of the firewall. If the firewall

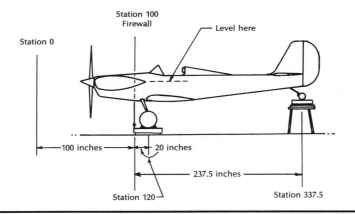

Fig. 2.5. Airframe reference system

is defined as station 100, then the main gear is at station 120 and the tailwheel is at station 337.5, as shown in Figure 2.5.

Now your can prepare to do the actual weighing. For this test the airplane must be fully equipped with oil and unusable fuel on board. All fairings, fittings, and canopies must be in place, even if they are only lightly screwed down. This test must also be done indoors, or in low winds if outside.

Use the most accurate scales available. Bathroom scales will give adequate results as long as a precaution is observed. Because the scale errors are unknown and may be significant, you must use the same scale to measure all three landing-gear loads. Then the error on all three locations will be similar and will have a reduced effect on any errors in your calculated c.g. As you will see when we look at the factors that affect stability, c.g. is a powerful tool to make your airplane handle well. Using the same scale will tend to minimize errors in the calculated empty c.g. while allowing a greater risk of error in the empty weight.

To begin the weighing, place the landing gear on the scale(s) and, using wood blocks, plates, stands, milk cartons, etc., level the aircraft both longitudinally and laterally. There are several ways of determining whether your airplane is level. Usually the plans will indicate a horizontal piece of structure or longeron accessible in the cockpit upon which a bubble level can be used. You can then level the airplane laterally by placing a level across the cockpit rim. Some larger airplanes use two points, one above the other, which can be brought into line using a plumb bob.

Another good way to level an airplane, especially laterally, is with a water level. This is a long transparent plastic tube almost filled with water that has a little food coloring added to make the water visible. Calibrate the level by holding both ends up together and marking the water level in both tubes with a grease pencil or a felt-tip pen. Carefully tape the ends of the tubes to the wing tips with the grease pencil marks at the same spot on each wing tip. When the water level matches the marks, the airplane is level laterally.

Once the aircraft is leveled, record the weight on the scale(s). Then remove the wheel from the scale(s) and weigh any chocks, leveling plates, etc., that you used on the scale for that wheel when it was weighed. This weight represents the zero reading for that scale and is called the tare weight. If you are weighing one wheel at a time, once you have recorded the measured and tare weights, move the scale to the next wheel and repeat the process. Record all the numbers on a form like that shown in Figure 2.6.

Once the weights are recorded for all three wheels, it is time to calculate the c.g. and weight for the empty airplane. On your c.g. form (Fig. 2.6), first subtract each tare from its corresponding measured weight to get the actual weight at each wheel. Now multiply each actual weight by its corresponding Station location to get the moment for each wheel. Add up all the moments as shown on the form and add up all the wheel loads to get the empty weight. Divide the sum of the moments by the empty weight to get the c.g. of the empty airplane. Empty weight does not normally include engine oil, so if the airplane was weighed with oil in the engine you should subtract the oil weight and moment as shown.

Now that you know the empty weight and c.g. of your airplane, what good is this information? At this point, find the figure for the safe c.g. envelope provided by the designer or by analysis if this is a home-designed airplane. You should use the empty weight/c.g. and compare it with the safe c.g. envelope to determine if you have any loading problems. One useful way of doing the comparison is to use what weights engineers call a potato diagram. As shown in Figure 2.7, it is a good way of visualizing the effect of adding or subtracting things from your airplane. Most light planes have a simple, rectangular gross weight/c.g. plot. Some large aircraft and helicopters have other factors that affect the diagram and yield a much odder shape, which some people (the weights engineers at least) feel resembles a potato. Thus the potato diagram.

To make the potato diagram, plot the gross weight and c.g. limits; these will usually appear to create a rectangle. Next calculate the effect on gross weight/c.g. of adding variable weights to the empty weight. Now plot these points and the empty weight/c.g. on the potato diagram. If any of the plotted conditions fall outside the envelope, you now have a picture of the problem and what is causing it. You should do this exercise for all the normal loading conditions you can think up. These should include heavy pilots, light pilots, full fuel, no fuel, light pilot with full fuel and baggage, etc.

If you have a gross weight/c.g. problem, there are several ways of coping with it. First, you can move parts of the airplane to adjust the empty c.g. A particularly useful item is the battery, which is heavy but relatively flexible in its location. If a major change is required, the engine can be moved by shortening or lengthening the engine mounts. This may be necessary if you have installed an engine significantly different from that called out in the plans. A third option is the addition of ballast. This

WEIGHT AND BALANCE FORM

Owner's Name ...John Brown...................... Aircraft N 5641 Z. Date 3-5-87......

Address ....4 Airport Drive.....................

.........John Wayne, Indiana..................

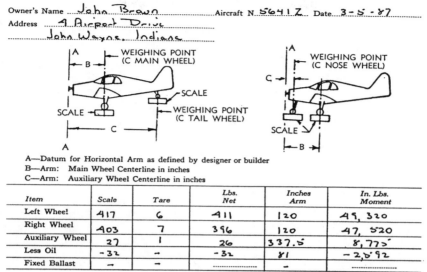

A—Datum for Horizontal Arm as defined by designer or builder
B—Arm: Main Wheel Centerline in inches
C—Arm: Auxiliary Wheel Centerline in inches

Item	Scale	Tare	Lbs. Net	Inches Arm	In. Lbs. Moment
Left Wheel!	417	6	411	120	49,320
Right Wheel	403	7	396	120	47,520
Auxiliary Wheel	27	1	26	337.5	8,775
Less Oil	-32	—	-32	81	-2,592
Fixed Ballast	—	—	....	—	....

Empty Weight   Total Moment...103,023...in. lbs.

$$\text{Empty C.G.} = \frac{\text{Total Moment}}{\text{Empty Weight}} = \frac{103,023}{801} = 128.6 \text{ inches}$$

FORWARD AND REARWARD CG EXTREMES

Item	Weight	Arm	Moment	Weight	Arm	Moment
Aircraft EW	801	128.6	103,023	801	128.6	103,023
Oil	32	81	+2,592	32	81	+2,592
Pilot	160	190	28,500	210	190	39,900
Passenger	210	153	32,130	—		—
Fuel	180	115	20,700	12	115	1380
Baggage	....	—	....	20	205	4100
Totals	1,373	136.2	186,945	1075	140.5	150,995

Forward CG   1 = ...136.0... in.   Rearward CG   1 = ...140.5... in.

Maximum allowable weight is: ......1400................ lbs. CG limits are

........136.0.........in. Forward CG, and ......140.5........in. Rearward CG

Equipment installed when weighed is as described in Aircraft Manual, Equipment List dated ....

...., except for the following items.

Item	Inches Arm	Lbs. Wt.	In. Lbs. Moment
....			
....			
....			

Fig. 2.6. Reprinted, by permission, from the EAA Service Manual *of the Experimental Aircraft Association*

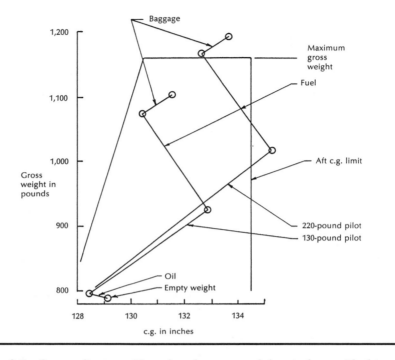

Fig. 2.7. Potato diagram. Note that the owner of the airplane with this potato diagram has a problem. A heavy pilot and no fuel will put this airplane outside the aft c.g. limit. This is especially dangerous because the pilot can take off within the acceptable envelope and by burning off fuel move outside the aft limit while in flight. This airplane may require a limitation on pilot weight.

is not a very elegant solution, but it provides a simple method of tuning the empty c.g. to make the airplane more useful. A small warning at this point: If you have made a major alteration to a design, such as a bigger engine, it is smart to weigh your airplane and plot the potato diagram before it is fully assembled, in case major alterations are required.

Rather than altering or ballasting the airplane, you may decide to live with the fact that certain loadings will exceed the limits of the potato diagram. This should not present a problem as long as these loadings are unlikely to occur or, conversely, are easy to avoid.

As you will find later on, c.g. is an extremely important tool in ensuring good ground handling, stability, and stall-spin characteristics. Your ability to understand the effect of c.g. on these parameters is going to depend strongly on your ability to know where your c.g. was located when certain tests were done. This requires not only a good determination of empty weight and c.g. but also a way of tracking it from flight to flight. The easy way to do this is to have a form, similar to the one shown in Figure 2.8, which you fill out and calculate for each test flight. You may even find out that you have to add ballast to achieve the desired c.g. for a given flight.

WEIGHT AND BALANCE CARD

Flight no.	4	Aircraft	56412
Date	4-28-87	Pilot	J. Brown
Target weight	1,130	Target c.g.	Aft.

	Weight (lbs)	Station (in.)	Moment (in.-lbs)
Empty weight	801	128.6	103,009
Engine oil	7	81	567
Fwd. seats	0	—	
Aft seats	185	190	35,150
Baggage	0	—	
Ballast	5	153	765
Zero fuel weight	998	139.8	139,491
Fuel	130	124	16,120
Takeoff weight	1,128	138.0	155,611

Fig. 2.8. Weight and balance card

While your aircraft is level for weighing is a good time to put attitude reference marks on the windshield. Turn the airplane toward a clear horizon, sit in the cockpit, and have a helper block up the tail until the airplane is level. With a grease pencil draw a horizontal line on the inside of the windscreen that matches the horizon from your eye position. Then have your helper lower the tail until the aircraft is 5 degrees nose-up and draw a short line on the horizon. Repeat the process at 10 and 15 degrees unless the horizon falls below the cowling. From the center of the zero line, draw lines downward at 30 and 60 degrees, as shown in Figure 2.9. You now have a tool that will enable you to fly pitch and roll attitudes with great accuracy as long as the horizon is visible through the windshield.

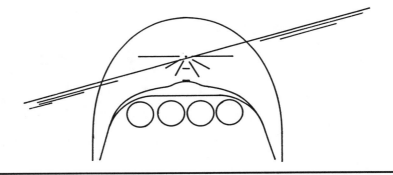

Fig. 2.9. Attitude reference marks

Landing Gear

The rigging, or alignment, of the landing gear is an especially critical check for taildraggers, since it has a significant effect on their ground stability. Get two square iron or aluminum bars and mark each with three lines at intervals of at least two feet. C-clamp one to the inside of each wheel's brake disk with the middle mark beneath the axle. Carefully measure between the bars at the forward and aft lines and determine the difference between the two distances. Unless the plans specify otherwise, the difference between the two measurements should be less than ¼ inch.

Realigning poorly rigged wheels can be difficult, depending on the type of gear. Tapered shims, which go between the leg and axle assembly, are available for Cessna-type spring gear. Oleos can be aligned by shimming or shaving the bearing surfaces of the scissors. Live-axle or split-axle gear, as found on the Piper Cub, are nearly impossible to change once they are built. Your best bet is to be extremely careful in the building phase.

Further Reading

Bangham, M. L. 1988. "A Learning Experience," *Sport Aviation* (February), 25.

Bingelis, Tony. 1972. "Alignment Procedures," *Sport Aviation* (March), 46.

_____. 1974. "How to Control Fuselage and Landing Gear Alignment," *Sport Aviation* (June), 54.

_____. 1986. "Maintaining Alignment during Construction," *Sport Aviation* (September), 29.

_____. 1987. "The Basic Fuel Pump System," *Sport Aviation* (August), 36.

_____. 1988. "Biplane Assembly and Rigging Procedures," *Sport Aviation* (February), 27.

_____. 1989. "Flight Testing Homebuilts, Stage One: Making Preparations for Flight Testing," *Sport Aviation* (January), 27.

Hall, Stan. 1986. "How to Move the CG," *Sport Aviation* (May), 28.

Hansen, Eric. 1987. "How to Test Your Pitot Static System . . . At Three Different Levels of Complexity," *Sport Aviation* (May), 37.

Hollman, Martin. 1986. "Structural Testing of the Lancair 200," *Sport Aviation* (January), 20.

Larson, Chuck. 1987. "Plastic Tube Alignment Guide," *Sport Aviation* (November), 26.

Murray, Chris. 1985a. "Weight and Balance . . . by Computer," *Sport Aviation* (February), 40.

_____. 1985b. "Weight and Balance . . . by Computer, Part II," *Sport Aviation* (March), 48.

Owen, Ben. 1985. "Fuel System Ground Check," *Sport Aviation* (December), 66.

_____. 1987a. "Wing Incidence," *Sport Aviation* (January), 60.

_____. 1987b. "Homebuilt Powerplant Precertification Inspections," *Sport Aviation* (February), 59.

Sequoia Aircraft Corp. 1987. *F.8L Falco Flight Test Guide.* Chapter 1, "Final Inspection." Obtainable from Sequoia Aircraft Corp., 2000 Tomlynn St., P.O. Box 6861, Richmond, VA 23230.

Thurston, David. 1987. "Powerplant Installation," *Sport Aviation* (May), 56.

Tunstall, Reginald. 1970. "Flight Testing the Homebuilt Airplane," *Sport Aviation* (January), 23.

[CHAPTER 3]

Final Preparations

At this point you are getting close to flying your airplane. If you are planning to test-fly your creation, it is time for you to prepare yourself. There are three areas you need to work on. First, you must develop the procedures required for contingencies, such as mechanical problems in flight. Second, you must bring your flying skills up to the level you'll need for handling your airplane. Third, you must develop the attitude or mind-set required to avoid problems and minimize risk if they occur.

Emergency Procedures

Developing emergency procedures consists of two parts—determining the procedures and learning how to do them in the cockpit. Start by listing every system failure you can think of. Then list the symptoms that would accompany each failure and what you would do to respond to each failure. What you are doing is actually writing the emergency procedures section of a flight manual. This is good mental exercise to do while commuting or eating lunch. An example of the output you should create is shown in Figure 3.1. After you are finished, ask other knowledgeable people to review your emergency procedures to see if they can find weak areas. As always, you must be the final judge of how adequately your procedures are written.

EMERGENCY PROCEDURES

Lateral Control Jam
Symptom: Stick locked in the lateral axis
Action: 1. Hold stick fixed against the jam and use pedal to remain upright and control heading. Use pedal inputs and shallow-bank turns to accomplish wide pattern. Land into the wind.
2. If stick is jammed far enough from neutral so that the aircraft cannot be held upright and/or heading cannot be maintained with pedal, relieve pressure against the jam momentarily. If jam clears, land as soon as possible and look for cause of jam. If jam persists, bail out.

Loss of Single Aileron Cable
Symptom: Reduced roll control. One aileron does not move in one direction when the stick is moved.
Action: Land as soon as possible, using shallow banks and wide pattern. Land into the wind to reduce roll-control requirements.

Loss of Electrical Power
Symptom: Loss of all electrical functions including radios, lights, electric turn and bank, fuel pump, flaps, etc.
Action: Master switch off. Return to (uncontrolled) field and execute no-flap landing.

Electrical Short
Symptom: Sparks or smoke in cockpit or cabin. Crackling on the radio or snapping heard in the cockpit. Erratic motion of electrically driven instruments.
Action: Master switch off. If smoke thins, return to field and execute no-flap landing. If smoke persists, land immediately.

Fig. 3.1. Emergency procedures example

Now for some procedures training, or as some people call it, hangar flying. Sit in the cockpit and practice the procedures you wrote. If it's winter, or if you plan to fly during the winter, wear full winter flying gear. This will not only teach you how to cope with most of the major systems problems you might have but it may also show you that some of the procedures aren't possible. If a switch or circuit breaker or valve cannot be reached when you have a heavy coat on, this is the time to find out and adjust for it.

If you have retractable landing gear, put the airplane up on jacks and verify its operation. If the gear has an emergency extension system, try it out too (with winter clothes on, if applicable).

Flying Skills

Most amateur builders neglect their flying skills while they spend three years or more building. Now that your airplane is almost ready to fly, it is time to face the most difficult decision you must make: Should you fly it, or should you get someone more capable to do the first few flights? To answer this you must review your experience and compare it with the probable characteristics of the airplane you've built. If you have 150 hours, all in Cessna 152s and 172s, and you haven't flown in a year, it is not smart to try to do the first flight on your Pitts S-1 with 180 horsepower and symmetrical wings. Normally, though, the situation is not this clear-cut. If you get someone to do your flying, remember that it's your airplane and the pilot must fly it the way you want it flown.

If you have decided to fly it yourself it is time to brush up on your basic skills. First get your biennial flight review squared away. Then put in a couple of hours practicing airwork in an airplane with characteristics as close to the characteristics of your homebuilt airplane as possible. If there are no taildraggers available and yours is, it is worth a trip to shoot some landings in an airplane with the little wheel on the back end.

In your practice flying, emphasize steep turns and spins if the airplane can do them. You should be striving to fly the airplane as accurately as possible. Now do some takeoffs, full-stop landings, and go-arounds. Again, emphasize precise airspeed control and accurate patterns. You should be doing this out of the same airport from which you intend to fly your airplane. While you are flying in the local area, learn the local landmarks and make sure you understand the boundaries of your restricted zone. Examine and memorize the locations of the most likely emergency landing spots with relation to the runways.

The final step comes after you have written your plan for the first flight. Get your practice airplane and go out and simulate the first flight from beginning to end. As you fly it, try to keep track of emergency landing areas and decide how you would get to them or to the airport in case of trouble.

Mental Attitude

You are now physically prepared to fly your airplane. Mental preparation requires something more. You have put a lot of time and effort and a lot of yourself into this creation. If something goes wrong, you will have a strong psychological urge to try and save your airplane, even if it means risking yourself. More than a few professional test pilots have died because they couldn't resist this tendency. You must convince yourself that the airplane is replaceable—you are not. You must be prepared to sacrifice the airplane to protect yourself if the worst should happen.

Developing this mind-set is not easy, but there are some techniques you can use to help. The most important is to think through the possibilities and rationally decide the best way to handle each one. For example, how would you respond to a complete engine failure after takeoff but before the crosswind turn? After the crosswind turn? How would you respond to a partial power failure during the crosswind turn? You didn't think of that one, did you? The objective here is to try to come up with decision gates that will help you make the right decision under pressure when there is a natural tendency for the mind to freeze, at least temporarily. Having a library of preanalyzed responses allows you to make a good decision and frees your mind to continue to analyze the situation.

A second thing you can do is to mentally review these actions when you do your practice test flight. As you do that practice flight, periodically ask yourself: How would I respond if the engine were to fail here? If you can't find a spot that you think you can make, plan to do your actual test flight at a higher altitude or a different place.

When you have finished this whole mental process, you will have conditioned yourself not just to fly the airplane but to test-fly it in the safest manner possible. You are now prepared to fly your airplane.

Ensuring the condition of your airplane does not stop, however, when you begin your taxi tests, nor when the FAA paperwork is done. An important part of any test program is an inspection intended to catch and correct mechanical problems before they put you at risk. In order to accomplish this, you will use two different types of inspections, depending upon the circumstances. These are the preflight and structural inspections.

Preflight Inspection

The preflight is similar to a normal lightplane preflight, and you should do it every time you start the engine with the intent of taking your airplane onto the runway. It consists of the following five steps.

1. Inspect the propeller for security and general condition.
2. Inspect the engine compartment for rubbing, leaks, and adequate oil level.
3. Check the landing gear for tire condition, brake condition, and leaks.
4. Check all functions of the control system for travel, lack of binding, strange noises, etc.
5. Inspect the skin of the aircraft visually for any deformation that could indicate internal damage.

Structural Inspections

You should do a structural inspection every time a limit, such as V_{NE} or limit load factor, is reached; after any structural incident, such as a hard landing or inadvertent aerobatic maneuver; or after a significant increase in envelope, as described in Chapter 6. The whole purpose of the structural inspection is to uncover structural damage before it puts you or your airplane at risk.

To do the structural inspection, remove all fairings, cowlings, and inspection plates. Inspect the engine mounts, both tubes and weldments, for any evidence of cracking, such as cracked, chipped, or discolored finish. Use a good flashlight, a mirror, and a magnifying glass and look at all the segments of the mount from all sides. Check the bolts, washers, and flanges for any evidence of bending or slippage.

If a crack is suspected but not obvious, the only good method to verify its existence is the dye-penetrant technique. Using this technique, you spray a low surface-tension dye on the affected component after removing dirt and the paint or finish. Then wipe off the dye and spray on a developing agent. The developing agent acts like a sponge, soaking up dye remaining in cracks so small as to be barely visible with a magnifying glass. The dye reacts with the agent to form a bright stain. An A & P or IA can do a dye-penetrant check of a suspect area for you. There is also a relatively cheap kit available. This type of test does demand technique, so practice on some scrap until you feel comfortable with the technique.

After inspecting the engine mounts, check the rest of the engine compartment for leaks, rubbing, and cracked or damaged baffling.

Inspect the fuselage structure using the same methods you used for inspecting the engine mounts. Pay particular attention to the structure adjacent to load-carrying fittings, such as that near the engine mounts, landing gear, wing, seat, and tail attachments. Look for any evidence of shifting, deformation, or cracking. In wooden structures be alert for any discoloration in the wood or on the surface around bolted metal fittings. It could indicate rot within the wood or a bolt rusting from exposure to moisture in the wood.

Inspect the wings, paying particular attention to the wing and control-surface attachment fittings. Look for evidence of shifting due to hole elongation from overload. Have someone shake the wing at the tip while you watch the fittings for evidence of slop. Use your fingers and ears as well as your eyes. Sometimes motion small enough to be difficult to see can be felt or heard as a popping or clunking sound. Control system linkages and hinges should also show no evidence of slop or deformation.

One method of checking a wing structure for damage is a natural frequency check. Grab the wing at the tip and shake it vertically, matching your inputs to the response frequency of the wing. When you have a stable response, count the number of cycles in a minute using the sweep second hand of your watch and record this number. This is the natural frequency of the wing in cycles per minute, which is dependent on the wing mass distribution and the wing spring rate. If there is significant structural damage to the wing, the natural frequency will probably decrease. This is not an absolute test, but if you ever detect a frequency change, look for structural trouble within the wing. This test works particularly well with long flexible wings such as those on gliders. However, it may be difficult or impossible to do with very stiff wings or with a biplane whose natural frequency is too high to be accurately measured by hand. Because the mass distribution of the wing affects the natural frequency, this test must be done with no fuel in any tanks in the wing.

The methods mentioned up till now work well with wood, steel-tube, or aluminum structures. Composite structures present a whole different group of problems. For these structures, the natural frequency check is a good first indication of trouble, but it is not definitive and it does not show where the damage is.

Damage in composites normally appears as either buckling or delamination. Buckling is a breakdown, due to overload, of the polyester or epoxy matrix surrounding the fibers. Because the highest stresses normally occur at the outer surface of a structure, and because in composites the skin is normally a part of the primary structure, this type of failure usually causes damage you can detect with external inspection. It will appear as loss of luster or visible crumbling of the skin. If there is any question, strip the paint off the affected area and look for a change in color. A damaged area will appear cloudy instead of translucent, and the fibers may be visible within the matrix.

Delamination is a failure of the bond between two layers of glass. It is normally caused by poor building techniques, but because it can be encouraged by fatigue it can appear anytime in the life of the airplane. Sometimes it can be felt as a soft area, but often it gives no surface indication. Fortunately, there is a simple technique called coin tapping that you can use to test for it.

Gently tap the area to be inspected with a quarter and listen to the sound. If the structure is solid, the response will be a sharp click. If the structure has delaminated, the sound will be a dull thud. With practice, you can actually map out the delaminated area on the surface of the wing. This technique works best in areas of primary structure, such as the spars, which are also the areas of primary concern.

If any of the above inspections or tests uncover structural problems, it is not enough to simply repair the area and return it to its original strength. If the failure occurs during envelope expansion up to and including normal limits, it may indicate a poorly made part or a design flaw. First, carefully compare the failed component with the drawing to ensure that there is no basic construction error. If there is no apparent mistake, contact the designer and describe your problem accurately and in detail. You may find that this is a known problem and that a fix is already available, or you may be the first person to report a problem. In any case, the area must be repaired so that it is stronger than it was originally. Repair techniques for wood, steel, and aluminum structures are well established. Glass structural repairs are not so simple. In the case of repair of glass structures, coordinate the method and the materials with the designer.

[CHAPTER 4]

Taxi Tests

At this point, your airplane is ready and you are ready. It is time for your homebuilt to move under its own power. The purpose of taxi tests is to prepare you and the airplane for that first trip "over the fence." You should proceed from slow taxi to fast taxi to land-backs, using a progressive, step-by-step process. I have laid out the process so that each step prepares you for the following step. It helps develop the appropriate piloting skills and is designed to identify and correct problems that might make the next step difficult or risky. It is also designed to provide you with a fallback position for each step. If at some point you develop a serious problem, you can fall back to a maneuver or portion of the envelope that you know you and the airplane can handle.

Some people believe that taxi testing is unnecessary and excessively dangerous. Their usual argument is that they want to reduce their exposure by quickly getting away from the ground. A large proportion of homebuilt accidents, though, are caused by loss of control through instability, inadequate control power, or pilot-induced oscillation. In addition, tailwheel airplanes have significantly poorer control on the ground when power-off than when power-on. Most of these deficiencies are detectable during taxi tests and land-backs, allowing you to fix the problem before proceeding to altitudes or speeds at which the penalty for loss of control is much greater. Putting it another way, would you rather find out your airplane is severely unstable and has inadequate yaw control at 30 knots on the runway or at 60 knots while landing from your first flight?

During fast taxi tests or land-backs, there are several situations that might force you to fly prematurely. For this reason, you must be prepared to fly whenever you taxi onto the runway. The airplane must be complete, with cowling, spinner, and fairings attached and safetied. It is smart to leave wheel pants off, however, to allow for easier tire inspections and brake cooling. The FAA preflight inspection must be completed and the airplane signed off for flight. The aircraft should be loaded to a c.g. between mid and forward for best handling qualities. You must be wearing a parachute, helmet, and harness and have practiced emergency bail-out procedures.

The first time you taxi under power, use wing walkers, especially if you intend to start on a crowded ramp. This is because the first problem you are likely to encounter is weak brakes. The way to discover them is not by dinging someone else's plane. If you have poor brakes, first check that they have been properly bled. Also check the brake master-cylinder geometry (Fig. 4.1) to ensure that you are getting adequate cylinder travel.

Another common problem in homebuilts is rusting of the brake disks. The steel disks are exposed to the weather, and if they are not used frequently, they will accumulate rust. The resulting pitting of the disks reduces the effective brake area and can cause a dramatic loss of

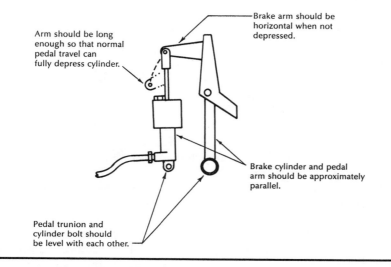

Fig. 4.1. Normal hydraulic brake pedal geometry

braking capability. The solution for this problem is the use of chromed disks. The plating prevents rusting and allows maximum contact between the pucks and the disks.

From this point on we will examine the taxi process by using an example. Assume that your airplane is a 150 HP taildragger with a fixed-pitch prop and an estimated stall speed of 45 knots. Although your airplane may differ, the techniques we will be discussing are universal.

When you taxi out for each run, consciously pick an abort point on the runway. This should be a point from which you know you can stop the airplane on the remaining runway at the speeds you will be achieving. This is the point at which you must be firmly on the ground, power off, and ready for braking. If you are still in the air at the abort point, smoothly apply full power and execute your first flight, following the flight card you have attached to your panel (we will discuss this in Chapter 5). Don't cut the abort point too fine, as the object is to avoid a flight if at all possible. You will probably adjust the abort point as your speeds increase and your experience with the airplane grows. For our example, the initial abort point should allow at least 1,000 feet of stopping distance.

First Taxi

For the first run, release the brakes and smoothly apply run-up power (about 1,500 rpm). The airspeed system should start reading at or before 40 knots, so allow the airplane to accelerate, with the stick back, to the abort point or 40 knots, whichever comes first. Smoothly retard the throttle and brake to a stop.

If this exercise was comfortable and you had no problem holding the airplane on the centerline, repeat the run using 200 more rpm. As long as you're comfortable, keep repeating the process until you can apply full power, accelerate to 40 knots, and stop comfortably.

This may seem a simple exercise, and it is for a tricycle gear airplane. For a tailwheel airplane, such as in our example, getting through this phase can be time-consuming, frustrating, and dangerous. This is because of a number of factors, including the inherent instability of tailwheel airplanes, reduced control at low speeds, and the lack of time and space to recover from a loss of control. In addition, control on the ground is a highly dynamic situation, so time of response and pilot-induced oscillation (referred to as PIO) are also factors.

Before we examine the problems that can arise during taxi tests and their cures, it is important to understand why tailwheel airplanes are unstable on the ground and why they can be so difficult to control at low speeds.

For a tailwheel airplane to balance on all three wheels, the c.g. must be aft of the main landing gear. If any side force is generated, as in a turn or a crabbed landing, the resulting centrifugal force acts at the c.g. Because the c.g. is aft of the main wheels, as shown in Figure 4.2, this force will tend to push the tail out or tighten up the turn. The forces the pilot can use to oppose this tendency are those generated by the rudder, tailwheel, and differential braking. At the low speeds we are discussing and with the power off, airflow over the rudder will be minimal. This will

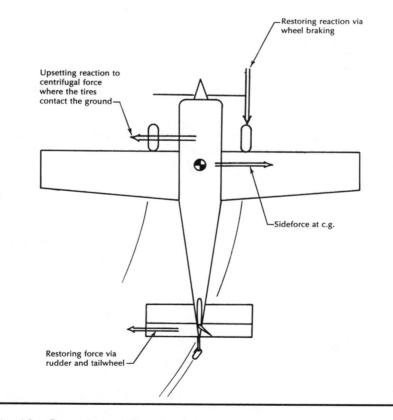

Fig. 4.2. Ground instability of tailwheel airplanes

make the rudder pretty ineffective, leaving the tailwheel and brakes for control. If the turn is allowed to become too tight, the side force can become so large that full tailwheel deflection and differential braking cannot prevent the classic ground loop, in which the airplane spirals into a tighter and tighter turn until a wingtip hits the ground, the gear fails, or the airplane stops.

As you can see, there are two problems involved: the basic instability of the airplane and the forces the pilot can generate to maintain control. Other than tightening the tailwheel springs and verifying the effectiveness of the brakes, there is very little the pilot can do to increase control. A more effective way of addressing the problem is to reduce the airplane's basic instability.

Instability Problems

Stability in low-speed taxi is predominantly affected by two factors: the distance the c.g. is aft of the landing gear and the main gear rigging. As the c.g. moves aft, it creates an increased turning moment, making the airplane more unstable. It also decreases the tendency of the tailwheel to rise during heavy braking. As a crude rule of thumb, the tailwheel load should be about 15 percent of the gross weight when the airplane is ready to fly. If you suspect excessive aft c.g., you may have to move the main wheels aft. If you have Cessna-type spring gear, this can be done by inserting a steel plate between the gear and the wheel, as shown in Figure 4.3.

As mentioned above, a c.g. located far aft of the landing gear decreases the tendency of the tail to rise during heavy braking or full-power run-ups. Some people erroneously think this gives improved braking or steering. In reality, braking force is proportional to the load on the braked wheels. Thus, as the load on the tailwheel increases, the load on the main wheels decreases, and the braking force available before the wheels slide is reduced. At the same time, the increased load on the tailwheel makes steering more difficult, and in an extreme case it may flex the tailwheel spring enough so that the tailwheel trunnion actually tilts forward, making steering almost impossible.

Stability in low-speed taxi is also affected by landing gear rigging. Most airplanes are rigged with the wheels aligned with the aircraft's centerline. Any amount of toe-out will make your airplane much more

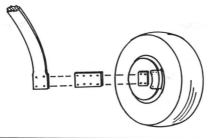

Fig. 4.3. Moving tires forward or aft

unstable. Small amounts of toe-in can be used to increase stability. However, this comes at the expense of increased tire wear and longer takeoff runs.

Although some people use toe-in to improve taxi stability, moving the main gear aft is by far the most effective method of improving taxi stability. As a general rule, unless your airplane is extremely powerful, you should be almost able to lift the tailwheel during a full-power run-up. A simple way to check the weight on the tailwheel, both statically and during a run-up, is to chock the main wheels and put a bathroom scale under the tailwheel. While you are in the cockpit, have someone compare the scale loads with the engine shut off and during a full-power run-up.

We normally think of the instability of a tailwheel airplane in terms of the pilot's ability to control a swing or the beginning of a ground loop. The phenomenon called pilot-induced oscillation (PIO), which is encouraged by simple instability, is something you may encounter for the first time during taxi tests.

If something like a gust of wind or a bump in the runway disturbs the airplane, it takes a measurable period of time for the pilot to respond. If this lag in response is sufficient, the pilot's response may be too large and too late, and it may cause the airplane to respond in the opposite direction. The pilot may then give another too-large-and-too-late response in that direction, and so on. The result is a back-and-forth instability in which the pilot adds inputs that keep making it worse. You will recognize PIO when the airplane is swinging back and forth and you are frantically responding but you just can't seem to get ahead of it. The most effective response to PIO in flight, when you have room and time, is to release or freeze the controls and then recover from the resulting maneuver. If you feel you are losing control of your airplane on the ground, this is not so easy to do. First, if freezing or releasing the controls is not practical, try to keep your inputs as small as possible. Immediately pull off the power, use the brakes if possible, and let the airplane stop.

The presence of PIO indicates that the natural frequency of response of the airplane is the same or close to that of the pilot. The pilot's natural response frequency is affected by training (you can learn to handle it), age, lack of sleep, tension, or anything that affects alertness. A positive fix usually requires changing the airplane's natural frequency.

During your initial taxi tests you may discover one problem that tends to promote PIO—a disconcerting tendency to roll away from a yaw (right yaw, left roll). For example, let's assume a gust of wind causes your airplane to swerve to the right. As it swerves, it rolls to the left. You counter with left pedal, causing the airplane to swerve left and roll right. This combining of yawing and opposite roll tends to be disconcerting enough to encourage the pilot to overcontrol, which can lead to PIO and eventual loss of control. At the very least it makes the airplane a lot less comfortable during takeoff and landing.

This trait normally results from excessively soft landing gear. It can also be made worse by a large mass in the wings, such as full tip tanks or wing tanks. Unfortunately, the only way to solve this problem is to stiffen the landing gear. If you have a composite gear, it can often be stiffened by adding more layers of glass. If you have a spring gear, it needs a thicker spring.

Repeat Taxis

Once low-speed taxi is under control, it is time to go faster and lift the tail. Use full throttle to accelerate as before to 40 knots with the stick back. When you reach your target speed, reduce power to idle and smoothly push the stick forward to lift the tail. The tail should lift cleanly with no yaw transients or big pedal inputs, and then it should settle back to the runway. If there are no problems, repeat the run, reducing the throttle about halfway before lifting the tail and then reducing the throttle the rest of the way when the tail is up. The last step is to use full power while you lift the tail and then reduce power at about 45 knots. During all of this, control should be adequate and there should be no large or violent control motions required to control the airplane.

The next step is to push the speed up to lift-off speed or 1.3V$_S$, as described in Chapter 5. For our example, this would be 58.5 knots; let's round it off to 60. With full throttle accelerate to 40 knots, lift the tail, and continue to accelerate. At 50 knots remove the power, allow the tail to come down, and stop. As before, no violent or large control motions should be required to control the airplane. Repeat this several times, incrementally building the top speed to 1.3V$_S$ (60 knots).

Use one of these runs to check the effectiveness and direction of the ailerons with a gentle roll reversal. The roll reversal, which is described more fully in Chapter 7, is a smooth, almost sinuous movement of the stick, first left, then right, then back to the trim position. Because aileron effectiveness is an unknown at this point and most ailerons generate a turning tendency (adverse yaw), this reversal should be very mild. Make it just big enough to rock the airplane slightly on the gear. If it seems silly to check the direction of aileron response, don't laugh; more than one production airplane has flown with the ailerons or some other control rigged backwards.

Land-backs

You are now ready to try land-backs. A land-back simulates a takeoff and landing and is the final step before first flight. Accelerate to 1.3V$_S$ (60 knots in our example) with the tail up, reduce power to idle, and smoothly bring the stick back. The airplane should rotate and lift off, even if momentarily. If the time in the air is very short (less than 5 seconds) leave a small amount of power on to stretch the time in the air during subsequent runs.

These power-off lift-offs will give you a first indication of your airplane's stability. The two problems most likely to appear at this point are wing heaviness and poor yaw stability.

Wing heaviness, which is normally caused by improper wing or aileron rigging, will appear as a tendency to roll right or left immediately after lift-off. Check the wing and aileron rigging to ensure that both wings and ailerons are symmetrical. Wing heaviness can be corrected by changing the wing incidence or twist, if you have a strut-based wing. If the wing is cantilevered, the simplest solution is to add fixed trim tabs to the ailerons.

Inadequate yaw stability will appear as a kind of snaking, a difficulty in keeping the airplane aligned with the runway. It is normally caused by inadequate tail area or excessive aft c.g.

Both of these problems must be fixed before you go any further. With most normal configurations, they tend to get worse with increasing speed and power.

At this point you are using up quite a bit of runway, so be careful. If you reach a point where you are having trouble getting the airplane firmly on the ground by the abort point or stopping it on the runway, it is time to consider skipping to Chapter 5 and doing your first flight. The risk in doing this is that you may discover undesirable handling qualities due to high engine power when you are in the air and committed to flight. If these are severe, they may force you to reduce power, possibly making an off-field landing. Only you can assess this risk against the risk of going off the runway during higher-power land-backs. The best solution, of course, is to do your testing on a runway long enough so that this doesn't become an issue.

If runway length is no problem and the power-off land-backs are comfortable, modify the procedure by reducing power to about half-throttle before lifting off and then reducing the throttle in the air to idle for landing. This will give you not only more time in the air but also your first indication of the effects of power on stability and trim.

In most conventional tractor airplanes, increased power reduces stability and causes noticeable changes in trim. Any stability condition that was marginal during the power-off land-backs will be worse with half power and worse yet with full power. Trim changes when you are reducing power to idle should be small, and there should be no evidence of wing heaviness, yaw instability, or the need for large control deflections to keep control of the airplane.

If your airplane is a pusher, be aware that pushers normally get more stable with increasing power. Thus, if your pusher had marginal stability in the power-off land-backs, you can probably expect it to get better with increasing power.

While doing these partial-power land-backs, repeat the gentle roll reversal you did while on the ground. It will give you a feel for both control sensitivity and adverse yaw. Adverse yaw is the tendency for an airplane to yaw away from a roll input (right stick, left yaw). It is an inherent characteristic of the aileron design and creates what is often called a rudder airplane. It is not dangerous, but you should be aware that roll control inputs will also require larger pedal inputs than you may be used to, just to keep the airplane coordinated.

A gentle roll reversal while in the air will give you your first real indication of how sensitive the roll control is. If small stick deflections create what seem like large responses, or if you have trouble stabilizing the airplane in roll after the reversal, you may have excessively sensitive controls. In the most extreme case, this can lead to PIO. If PIO occurs, you must immediately neutralize the stick and chop the throttle, allowing the airplane to flop back onto the runway. It won't be graceful, but it's a lot gentler than digging in a wingtip or sailing off the end of the runway still trying to gain control.

Excessive control sensitivity can be cured by reducing the gearing between the stick and the ailerons so that the same stick displacement gives less aileron response. Often you can accomplish this by simply drilling a new hole in the input side of the aileron bellcranks to create a smaller arm, as shown in Figure 4.4. You can easily return the ailerons to their old gearing if you decide this is advisable later on.

Sluggish or inadequate roll response is not as much of a problem as excessive response because the ailerons will get more powerful as the airplane goes faster. If, however, you find yourself using large inputs just to keep the wings level, or if you must use nearly full control to lift a low wing, roll control must be increased. Methods to do this are discussed in Chapter 10.

The final step in the taxi test is full-power land-backs. These should be attempted only if you feel very comfortable with the airplane at this point and you are having no trouble getting the airplane stopped on the available runway. The full-power land-back will give you a good feel for

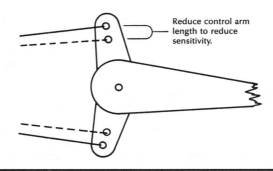

Fig. 4.4. Reducing control sensitivity

the trim changes associated with full power and will simulate a takeoff and a landing.

Apply full power, accelerate to $1.3V_S$ (60 knots in our example), rotate, and lift off. Level off about 3 feet above the runway. Then immediately and smoothly reduce power to idle and land. If your airplane has flaps, you should do the partial- and full-power land-backs with the flap setting you intend to use for your first couple of flights. If your flaps are particularly powerful, do not use more than half flaps for several flights.

During a full-power land-back you will be experiencing large, power-induced trim changes near the ground. As with the other maneuvers, you should be able to control your airplane without the use of large, violent control motions. The key to doing this maneuver is smoothness. Control inputs may need to be aggressive but they must also be smooth, with no jerky or violent inputs that cause violent aircraft responses. When reducing the throttle, smoothly reduce the power to idle. Don't try to hold the aircraft precisely at altitude and height. If the aircraft starts to deviate, use just enough control to stop the deviation; then smoothly bring the airplane back.

During land-backs, it is likely that you will experience some hard landings. Because landing loads are more abrupt than normal flight loads, they tend to feel deceptively mild to the pilot and indicate low on an accelerometer. If these indicate 1.5 g's or more on the accelerometer you installed, you must do a structural inspection. In addition to checking the normal areas of landing gear, wing fittings, and engine mounts, pay special attention to wing- and tip-tank fittings if the tanks contained an appreciable amount of fuel.

Tricycle-Gear Airplanes

Up till now we have used a tailwheel airplane as an example, since it requires the most complex taxi test procedure. This is because it is inherently unstable and because it requires a dramatic pitch-attitude change during acceleration. In general, other landing gear layouts require a less complex approach. For example, the sequence for a tricycle-gear airplane would simply involve building the taxi speed in increments to $1.3V_S$ and then doing power-off, partial-power, and full-power land-backs.

Although tricycle-gear airplanes are easier to taxi because they are stable on the ground, there are some problems you can have with nose gears. These are nosewheel shimmy and the inability to rotate.

Nosewheel shimmy is a side-to-side vibration of the nosewheel caused by inadequate nosewheel centering. It normally varies with speed and nosewheel load. If you encounter a vibration or shudder in the nosewheel, cut the throttle and come full back on the stick to reduce the load on the nose gear. If the shudder persists, get on the brakes and stop fast. Shimmy loads can be large enough to damage the nose gear.

There are several fixes for shimmy; unfortunately none of them is easy. You can add a hydraulic damper, as Cessna does on the 152 and 172. Besides adding complexity, though, the damper is a maintenance headache. A simpler solution is to increase the centering tendency of the nose gear. If you have an oleo or telescoping strut, you can increase the centering tendency by tilting the strut farther forward, as shown in Figure 4.5. A castoring gear can be made more stable by moving the wheel further aft of the castoring axis. An interesting fact is that double wheels are resistant to shimmy. If none of the other solutions intrigues you, try replacing your single nosewheel with a double nosewheel.

If you find you have to accelerate to a speed much above V_S and need to use a large stick deflection before you can lift the nose, you are

Increase nose-gear inclination.

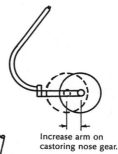

Increase arm on castoring nose gear.

Use double-tired gear.

Fig. 4.5. Fixes for nosewheel shimmy

having difficulty rotating. As your airplane accelerates down the runway, the drag of the landing gear and the weight of the airplane forward of the main gear both tend to hold the nose down, as shown in Figure 4.6. For the airplane to rotate, the moments generated by these loads must be exceeded by the moment generated by the download on the tail (or the upload on a canard). If the c.g. is far forward of the main gear, or if the main gear is very long and/or the tail cannot generate enough moment because it is too small or the tailcone too short, you may not be able to raise the nose. Even if you can raise the nose and the airplane lifts off, the moment caused by the c.g.'s being forward of the main gear will be relieved as the airplane's weight comes off the gear and the airplane will pitch up. This will require a sharp nose-down input from the pilot just after lift-off, a maneuver that can be disconcerting or even dangerous.

The most obvious solution is to move the main gear forward. If this is impractical, a partial solution is to lengthen the nose gear so that the aircraft sits at a more nose-up attitude on the ground. As you accelerate, the wing will start to generate lift, reducing the nose-down moment caused by the weight of the aircraft forward of the gear. The penalty of this option is that the wing will begin generating lift earlier in the ground roll, and the accompanying induced drag will increase the length of takeoff roll.

Another option is to enlarge the elevators to generate sufficient downward force to rotate at a lower speed. Although this does work and has the side effect of improving pitch stability, it does not relieve the pitch-up problem. It also adds weight and drag and increases the stabilizer bending loads. Don't try this unless you have the analytical capability and can avoid structural problems. Don't even consider

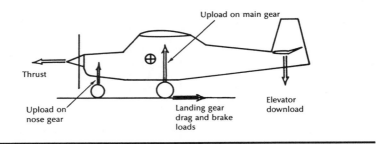

Fig. 4.6. Loads that affect ability to rotate

enlarging a canard. Because the canard is strongly destabilizing, enlarging it will probably mess up your pitch stability.

Another problem that can arise in airplanes with odd powerplant locations is the effect of thrust on pitching moment. As shown in Figure 4.6, thrust normally has little effect on rotational ability because the thrust line passes near the c.g. If the thrust line is far above the c.g., however, it will apply a nose-down moment that will make rotation difficult. For this reason, flying boats with pod-mounted engines sometimes require oversized elevators for adequate rotation ability. The problem is made more complex if the elevators are affected by thrust because they are located in the prop wash. This will make the elevators more effective when full power is applied. For example, if you have an airplane with a high thrust line (a pod-mounted engine), thrust will tend to push the nose down. Reducing the power will relieve this nose-down moment and cause the airplane to nose up, perhaps quite sharply. If the elevators are buried in the prop wash and are up, then chopping the throttle will reduce the download on the elevators and alleviate the nose-up pitching moment associated with the reduction in thrust.

The most important thing to recognize is that power changes in unconventional designs can be both unpredictable and violent. Think through very carefully the mechanics of your airplane and try to estimate ahead of time what power changes will do to your airplane. When you actually make the changes make them gradually and be ready for surprises.

Gliders

Although the basic ideas are the same, doing land-backs in a glider will require some special adjustments to the basic technique. The most effective way to do taxi tests with a glider is to do low auto-tows. Besides a car and a towline, you will need three helpers: one to drive the car, one to sit in the car and keep the driver informed about what you are doing, and one to handle the wing and watch for conflicting traffic. The tests should be performed early in the morning to avoid winds and traffic.

There are two methods of auto-testing a glider, depending on runway availability. If the runway has wide, smooth shoulders, the driver should first accelerate to a speed you have specified and should then, at a previously determined point or at cable release, pull off the runway before slowing down and stopping. This method is preferable because the car will be behind the glider during the landing and roll-out phase in case of control or braking problems. Moreover, if there is a cable-release failure, the cable should automatically back-release; this requires that the tow car be able to get well clear of the runway at speeds of up to 60 mph.

If the runway shoulders are not that good or the runway is especially long, the tow car can continue to the end of the runway. In that case, though, a cable-release failure does present some problems. You and the observer must have an agreed-upon point at which you should have released the tow cable. If you don't release by that point, the observer will inform the driver, who will slow down gradually, allowing you to land and stop behind the tow car.

The first tow should be up to a speed sufficient to allow good control in pitch, roll, and yaw (about 30 mph). This will verify the proper functioning of all three controls and give an indication of braking ability. On subsequent tows, build up the speed in increments until you can take off, accelerate to aero-tow speed (about 60 mph), release, land, and stop without problems. You should be able to do this without large or violent control deflections or large deviations from the runway centerline. If any of these occur, don't increase speed until you fix the problem. If at any point you are having real difficulty keeping control while on tow, release and land immediately. A ground loop or loss of control on touchdown is far preferable to an out-of-control impact with the ground while on tow.

If the runway length is sufficient, you may be able to try gentle pitch, roll, and yaw reversals (after releasing) to determine control authority and coupling effects, as described in Chapter 7.

Further Reading

Sequoia Aircraft Corporation. 1987. *F.8L Falco Flight Test Guide.* Chapter 2, "Flight Testing." Obtainable from Sequoia Aircraft Corp., 2000 Tomlynn St., P.O. Box 6861, Richmond, VA 23230.

[CHAPTER 5]

First Flight

At this point you've checked out all the aircraft systems critical to flight. You've practiced takeoffs and landings, you've verified control effectiveness, you understand how trim changes with power, and you have some idea of stability. The first flight should be anticlimactic, and that's the whole idea.

You've developed the necessary skills in yourself and know how the airplane will perform, so far as you can, so all that remains is to plan what to do on that first flight. The purpose of the first flight is to open an envelope large enough to allow a safe pattern and landing. This envelope will then serve as the starting point for the development of the rest of your airplane's envelope, as described in Chapter 6.

Flight Card

Planning the first flight consists of completing what is known in the testing business as a flight card. The flight card is a written plan that describes what you intend to accomplish on a given flight. It goes with you in the cockpit and serves as a checklist to ensure that you do all you have planned and nothing more.

Figure 5.1 shows a typical flight card as written for our example airplane (150 horsepower, 45-knot V_S). Your flight card may differ because of differences in speed or other factors, but the approach should be the same. As before, use a step-by-step or building-block approach to minimize surprises and to enable you to back off to a condition you have already flown, in case a maneuver or condition causes trouble.

The card for a first flight should be organized around a baseline speed; let's call it V_O. This will serve as your climbout speed, the starting point of your envelope, and the speed you fall back to in case of trouble. V_O should be equal to or greater than $1.4V_S$ and should be fast enough to allow good visibility and engine cooling. Conversely, it should be slow enough to allow a good rate of climb. For some airplanes with poor climb performance, this might mean climbing at the predicted best-rate-of-climb speed.

The flight card should also include a list of limitations. This list is to prevent you from doing something you shouldn't under the pressure of the moment. In Figure 5.1, for example, we had set out very strict limits on winds, c.g., and bank angle. These limits should be adhered to absolutely. Recheck your c.g. calculations before each flight. If the winds are over your limits, don't let anyone pressure you into flying. Only a major crisis should induce you to exceed a bank angle of more than 30 degrees below 1,500 feet.

The first item on your flight card should be something you've already done, in this case a full-power land-back. This will not only help you loosen up and shake out the mental cobwebs but will also indicate to you that the airplane is up to snuff.

After the land-back, taxi back and prepare to take off. As you sit on the end of the runway, run through a short mental exercise. Check control freedom and note the fuel level. Check the abort point you intend to use. Then imagine all the emergency landing spots for this direction of takeoff and decide which ones you will go for if the engine fails at 500 feet or at 1,000 feet.

First Flight Procedure

It's time to go. Apply full power and accelerate as you have done so many times before. Lift off and accelerate to V_O. Then pull the nose up and climb, maintaining V_O. In case of a problem, a landing on the runway is probably possible at or before V_O. Once you have started to climb, you are committed to an off-field landing.

Climb straight ahead at V_O to 1,000 feet and then execute a 180 climbing turn at a 30-degree angle of bank (AOB). This will put you in

FLIGHT CARD

Flight Number 1   **Fuel** full
Takeoff gross weight 1,230 lbs.   **C.G.** 129 in.

Limitations 10 kts. wind
           45 deg. maximum angle of bank
           30 deg. maximum bank below 1,500 ft. AGL
           100 kts. maximum speed

Maneuvers
1. Land-back from 80 kts
2. Takeoff
3. Climb, 80 kts.
4. LT in climb, 30 deg. AOB, to climbing downwind
5. RT, 30 deg. AOB
6. LF, 80 kts., 4,000 ft.
7. LT, 30 deg. AOB
8. LT, 45 deg. AOB
9. RT, 30 deg. AOB
10. RT, 45 deg. AOB
11. LF, 70 kts.
12. LF, 60 kts.
13. LT, 30 deg. AOB
14. RT, 30 deg. AOB
15. Stall, wings level
16. Stall, wings level
17. LF, 80 kts., ½ flaps
18. LT, 30 deg. AOB
19. LT, 45 deg. AOB
20. RT, 30 deg. AOB
21. RT, 45 deg. AOB
22. LF, 70 kts.
23. LF, 60 kts.
24. LT, 30 degs. AOB
25. RT, 30 deg. AOB
26. Stall, ½ flaps
27. LF, 80 kts., retract flaps
28. LF, 100 kts.
29. Descend, 80 kts., 500 fpm
30. Downwind, 80 kts., ½ flaps
31. Final, 75 kts., ½ flaps
32. Go-around, 80 kts., 0 flaps
33. Downwind, 80 kts., ½ flaps
34. Final, 75 kts., ½ flaps
35. Landing

Key LF = level flight; LT = left turn; RT = right turn; AOB = angle of bank

Fig. 5.1. Flight card

a position so that, in case of a problem, a simple power reduction will set up the downwind leg for landing. Check the engine gages at this point, with special attention to the cylinder head temperatures. If anything is at or in the red, simply reduce power, execute a normal pattern, and land. If all the gages are good, continue to climb, using 30-degree AOB turns for positioning to 4,000 feet above ground level (AGL).

There are several philosophies as to where to do the flying for your first flight. If the airport authorities have no objection, I suggest flying immediately over the airfield. My next choice would be upwind of the field. Both of these areas give you the best chance of getting back to the field in case of a major problem. In any case, your flying should be done at least 4,000 feet AGL over an unpopulated area.

At 4,000 feet, level off and reduce power to maintain level flight at V_O. There should be no large pressures on the stick or pedals, and the controls should be about at mid travel. Execute 30- and 45-degree AOB turns to both right and left. This is where the marks on your windscreen pay off. You should be able to fly pitch and roll attitudes accurately by placing the appropriate marks on the horizon.

Entering the turns should require only gentle roll and yaw inputs in the direction of the turn. Only minor longitudinal inputs should be required to hold altitude. The airplane should not pitch strongly up or down or have a strong tendency to roll into the turn.

Slow to 60 knots. The stick should come back a noticeable amount as you slow down. Again, do left and right turns but only to a 30-degree AOB.

Level your wings and very gently do one or more full stalls. Note the indicated stall speed (V_S) and any warning the airplane might be trying to give you. You should be able to hold the wings level with rudder, and there should be little or no wing drop at the stall.

If your airplane does not satisfy the criteria in the last three paragraphs, it is your first indication of a stability and/or a control problem. If the problem area is minor and easily controlled, continue the flight and fly the landing pattern with special care. If the problem appears to be severe, skip the flap-down portion of the flight, descend, and land. You will be using the techniques of Chapter 7 to analyze the problem and devise fixes. If there are no apparent problems, repeat the 80-knot (V_O) turns, 60-knot turns, and stalls with the flaps in the intended landing position. If any of the criteria appear appreciably worse with the flaps down, fly the pattern and land with zero flaps.

Descend at V_O and about 500 feet per minute (fpm) to pattern altitude. Opposite your touchdown point, reduce power, lower landing flaps, and slow to $1.4V_S$. On final, slow to $1.3V_S$.

If at this point everything is under control, you will want to do a missed approach. At about 100 feet smoothly apply full power, accelerate to V_O, retract the flaps, and climb at V_O. Execute the same pattern, using 30-degree bank turns and slowing to $1.4V_S$ on downwind and to $1.3V_S$ on final to the landing.

Follow-up

You will probably taxi in with the biggest grin you have ever had to find yourself showered with congratulations! Enjoy it—you've earned it. It's important, though, to resist the temptation to make another flight. You're not prepared for it, and you have other things to do first.

After the idolizing crowds have cleared out, sit down with a few knowledgeable friends and discuss the flight. In your mind, go back over everything you did and the numbers you saw, looking for any signs of trouble. Any potential problem needs to be fixed or to be investigated on the next flight.

The other thing that must be accomplished at this point is a full structural inspection. This is the first time your airplane has seen full-flight, aerodynamic, and vibration loads for any significant length of time. Any problems that show up now should be considered serious and should be dealt with promptly. Pay special attention to any evidence of rubbing or wear in the engine compartment or fairings. This kind of problem is simple to fix now and can cause a lot of aggravation if let go. Also, carefully check all fittings for slop or evidence of motion. Poor fits and joints with slop will show up more readily after a flight.

Further Reading

Bingelis, Tony. 1989. "Flight Testing Homebuilts, Stage Two: Making the Initial Flight Test," *Sport Aviation* (February), 27.

Chamberlain, Brad. 1985. "Dragonfly Fever," *Sport Aviation* (March), 51.

Galeazzi, Remo. 1989. "Steenship ... The Second Time Around," *Sport Aviation* (May), 52.

CHAPTER 6

Envelope Expansior

Any airplane's envelope is defined by a plot similar to that shown in Figure 6.1.This is a plot of load factor in g's versus speed, commonly referred to as a V-N diagram. Envelope expansion is the process of safely and methodically determining the envelope within which the airplane is structurally sound and free of flutter.

V-N Diagram

Because you will spend the next several flights probing the corners of this envelope, it is wise to examine the V-N diagram closely to understand the parameters that determine each line and point. The curved line on the left side represents the largest amount of lift (in terms of load factor) that the wing will develop at a given speed without stalling. As you can see, it starts at V_S (stall speed) at 1 g and increases in proportion to the square of the speed up to V_A. In fact, you can calculate this curve from your stall speed by the relationship $N = \left(\dfrac{V}{V_S}\right)^2$ where N is the load factor.

There is only one problem. All these speeds are in true airspeed (called TAS), and we don't have an airspeed system calibration at this point. Because most amateur-built airspeed systems indicate low at low speeds, a calculated V_A is normally lower than the real V_A, giving a conservative, and therefore safe, answer.

The positive-limit load factor, N+, is determined by the structural capability of the wing and tail and should be specified by the designer. FAR Part 23 indicates that N+ should be between 2.5 and 3.8 for normal category airplanes, 4.4 for utility category, and 6.0 for aerobatic category. I believe that no homebuilt should have a limit load factor less

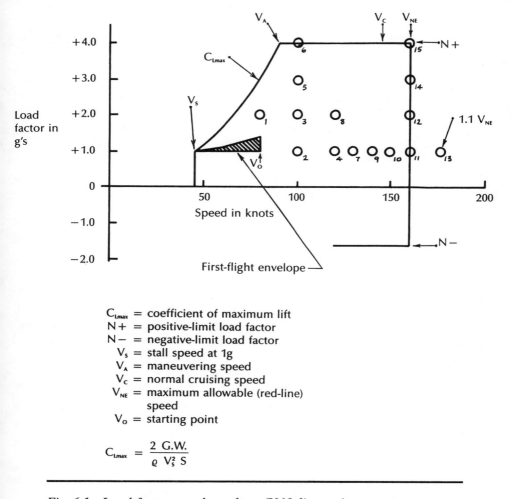

C_{Lmax} = coefficient of maximum lift
N+ = positive-limit load factor
N− = negative-limit load factor
V_s = stall speed at 1g
V_A = maneuvering speed
V_c = normal cruising speed
V_{NE} = maximum allowable (red-line) speed
V_o = starting point

$$C_{Lmax} = \frac{2 \; G.W.}{\varrho \; V_s^2 \; S}$$

Fig. 6.1. Load factor–speed envelope (V-N diagram)

than 4 because of the homebuilt's generally greater pitch response and lower wing loadings.

V_{NE} is the red-line speed beyond which you must not go during normal flight. It is defined as 90 percent of the highest speed demonstrated by test and can be determined by a wide variety of things. It may be determined by a structural limit, by handling qualities, or by the collapse strength of a windshield. It may even be an arbitrary number. In any case, it should be at least 15 percent greater than the maximum cruise speed (V_C, normally at 75 percent power).

As always, develop the envelope in a step-by-step fashion. The simplest technique is to build up in speed and then in load factor, with the last data point of the flight being the most strenuous. Because flutter can occur above speeds of about 140 knots, the level-flight points above speeds of 140 knots will include flutter tests in all three axes.

Figure 6.1 also shows an envelope development approach for our example airplane. The hatched area shows the envelope you developed during first flight. The sequence of points is shown by the numbers next to each point. Each time you achieve maximum load factor, you should do a structural inspection. The resulting test pattern will look like this.

Flight 2: Points 1 through 6
Structural Inspection

Flight 3: Points 7, 8, 9 (flutter), 10 (flutter), 11 (flutter), 12, 13 (flutter), 14, and 15
Structural Inspection

Although the FARs would require this envelope to be demonstrated over a matrix of weights, c.g.'s, and altitudes, this is not necessary for a homebuilt. For your purposes it is sufficient to do these points at light weight and heavy weight and one altitude. They need to be demonstrated at forward and aft c.g. at heavy weight only if your airplane has a large c.g. range. Most one- and two-place homebuilts have a small enough c.g. range that it is sufficient to do it at one c.g. location. Use some common sense here. If a normal maximum-gross-weight loading will put the c.g. at the forward limit, then this is where the heavy-weight points should be flown.

Flutter is a function of indicated airspeed (IAS) and aerodynamic damping. Engineers who deal in flutter use the term equivalent airspeed (EAS) to compare results of flutter tests with analysis. EAS is defined as follows:

$$EAS = TAS \sqrt{\frac{\rho}{\rho_O}}$$

where $\frac{\rho}{\rho_O}$ is the ratio of the air density at the actual altitude to the density at standard sea level. It is equivalent to IAS if the instrument and position errors are small. Because we are simply trying to determine that a particular airplane is flutter-free, IAS will work just fine. The use of EAS would be an unnecessary complication.

Aerodynamic damping decreases with altitude. For this reason, flutter normally becomes more critical at high altitudes. Once the full envelope is explored at low altitude, go to the maximum operating altitude and repeat the flutter points, starting with an indicated airspeed of 140 knots and gradually building up to $1.1V_{NE}$.

Two additional factors must be understood. Nothing says that only envelope expansion points can be flown on a given flight. There is no need to fly six points and then land. A more efficient method is to do some lower-risk testing, such as handling qualities work, at a point in the envelope that has already been developed and then finish the flight with the envelope expansion points. The second factor to keep in mind is that envelope expansion is high-risk flying. Flutter testing is especially risky and should be approached with a great deal of caution. You should always do this kind of testing at 5,000 feet AGL or higher and should wear a helmet and parachute. If you have an escape hatch or emergency door release, check it for proper functioning before each flight.

Flight Maneuvers

Now that we've reviewed what you are going to do, let's discuss how to do it. Load factors of up to 2 g's can be developed well enough in a coordinated turn. Higher load factors will require the use of a maneuver called a symmetrical pull-up. The technique used for the symmetrical pull-up is designed to avoid extreme attitudes both before and after the maneuver. To execute the pull-up, first trim for the target speed or the speed at which you want to achieve the desired load factor. (If full power in level flight won't get you to your target speed, then use a shallow dive. Power should always be used, in preference to diving, to achieve high speeds.) Next push over to some higher entry speed to begin the pull-up. Once you have achieved the entry speed, come smoothly but firmly back on the stick, trying to attain the target speed and load factor at about the same time. It will take some practice to determine the proper entry speed and how much stick force to apply, so creep up on the point by starting with a mild input and then repeating the maneuver using gradually more and more aggressive inputs. At no time should you allow the pitch attitude to exceed 30 degrees nose-up, even during a recovery. This is to avoid the possibility of a departure into a stall/spin or tail slide in an aircraft that has not completed spin testing.

Certain points on the V-N diagram may present some problems in terms of entry speed or the ability to develop the desired load factor without experiencing an accelerated stall. When you are doing the N+ point at V_A (point 6 in Figure 6.1), your entry speed must be sufficiently high to allow you to achieve N+ without an accelerated stall. If you are running out of aft stick or getting an accelerated stall before achieving N+, then increase the entry speed slightly. It is not necessary to hit the target speed on the nose. Achieving the target load factor is more important than doing it at exactly the right speed.

N+ at V_{NE} presents a slightly different problem because you do not want to attempt an entry speed greater than $1.1V_{NE}$. In this case you might have to settle for an N+ load factor at some speed slightly lower than V_{NE}. Again, the load factor is more important than the speed at which it is achieved.

The objective of envelope-expansion testing is to demonstrate the envelope without damaging the aircraft. Any change in noise, vibration, control feel, or anything else during or after a test point may be an indication that you have somehow damaged or changed the airplane. If this happens, you should immediately break off the test, slow down to a reduced speed, and return home with a minimum of maneuvering. Do a full structural inspection to look for any signs of damage or change. If the part that failed was improperly built (undersized, improper material, poor joint, damaged material, etc.), it may be sufficient to repair the part and build up to and repeat the test point. Improper building is not usually the case, however, because most homebuilders are meticulous. Moreover, the aircraft should have reserve strength (50 percent is standard practice) even at the demonstrated limit. There are two possible approaches to the problem. The simplest is to limit the airplane to a load factor that is two-thirds of that at which the failure occurred. The problem with this solution is that it may severely limit the utility of the airplane, especially if the failure occurred at a relatively low load factor. The more difficult solution is to redesign the affected part. This can be a very complex job. You should involve the original designer and, if you yourself are not technically capable, some professional help.

Flutter

The other test you will be doing during envelope expansion is testing for flutter. Flutter is a potentially violent phenomenon that occurs when an aerodynamic surface goes into resonance with another piece of the airplane. This mouthful doesn't really tell us how flutter happens, so let's look at an example. Figure 6.2 shows an end view of the airfoil and control surface as you would find them on a typical wing. You will notice that the center of mass of the control surface is aft of the hinge line. If a gust strikes the wing, it will displace it upward (b). Because the center of mass of the aileron is behind the hinge, it will tend to be left behind, or to lag downward as the wing moves up. The lagging control surface will feed a force into the wing that will twist the wing's leading edge down. This will give the wing a negative angle of attack and cause it to accelerate down (d) again, leaving the control surface lagging behind. Again the control surface will feed in an input, this time bigger, and the cycle is repeated. All physical things have a frequency at which they like to vibrate, their natural frequency. As you can see, the two systems (wing and aileron) must be able to feed energy to each other at about the same frequency for flutter to occur. Thus, the key to eliminating flutter is to prevent either the aileron or the wing from feeding vibratory energy to the other, to modify them so that their natural frequencies are different, or both.

Careful examination of Figure 6.2 will show that, in addition to the similar natural frequencies of the control surface and wing, several other conditions must be present for flutter to occur. The control system must be "soft" or sloppy enough for the control surface to move. The center of mass of the control surface must be aft of the hinge line to provide the control input. The wing must be torsionally soft enough for the wing to respond to the control inputs.

The control system's spring rate and the wing's torsional response are a direct result of the design, so it is not normally possible to change them. This leaves the softness of the control system and the mass distribution of the control surface as the best tools for the homebuilder to use to control flutter.

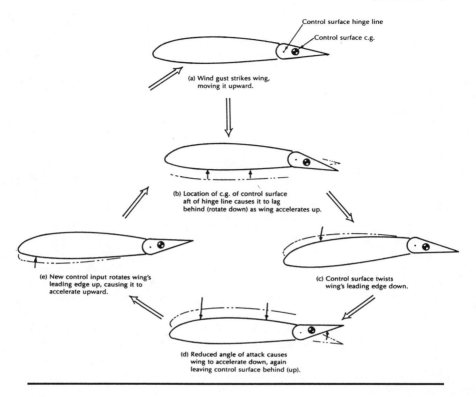

Fig. 6.2. How flutter works

As you have seen in Figure 6.2, the center of mass located aft of the hinge line is what feeds a moment into the control surface. Thus, the moment is reduced as the center of mass is moved forward closer to the hinge line. The process of moving the c.g. of the control surface closer to the hinge line is referred to as balancing, and if the c.g. coincides with the hinge line it is described as 100-percent mass balanced. Balancing a control surface is normally accomplished by adding weight forward of the hinge line.

Flutter can and does occur in any control surface. However, it is rare at low dynamic pressures (low airspeeds). If your airplane has a V_{NE} approaching or exceeding 140 knots, all control surfaces should be balanced. In addition, be careful to avoid slop or free play in your control system. The cardinal rule in flutter prevention is to avoid any changes that would reduce the stiffness of the control system or the torsional stiffness of the wing, stabilizer, or fin. Do not change the thickness of skins or spar webs. Do not change the skin area or replace stiff skin with fabric. If you are designing, make the wing as stiff in torsion as possible without excessive weight increase, and include provisions for surface balancing.

Now that we know what flutter is, let's see how to test for it. Accelerate to and trim at the target speed. Take your hands and feet off the controls and introduce a pulse into the desired control axis. Do this by firmly hitting the control with the palm of your hand or tapping a pedal with one foot. Only one pulse is required in each axis, and you should wait at least five seconds after each pulse to ensure that the axis is flutter-free. If you feel or see any vibration, either in the control system or through the airframe, immediately grab the stick and smoothly pull the nose up to reduce speed. Remember that flutter is encouraged by high speed and a soft control system. Simply grabbing the controls and applying a force to them will stiffen the system and may eliminate the flutter. Pulling up will reduce speed, which will also help to get you out of trouble. This is one type of flight testing where things can happen very quickly, so at the least sign of vibration, act fast!

Flutter gets worse with altitude at the same indicated airspeed. Thus, the worst condition is $1.1V_{NE}$ at maximum altitude. You should approach this point by building up in small increments of speed at a moderate altitude (5,000 ft AGL). Then repeat the process at your maximum test altitude.

Any repair, repainting, or physical change to a control surface will alter its balance and thus its tendency to flutter. Control surfaces must be balanced after painting. If any work is done on a control surface, it must be removed and rebalanced.

Further Reading

Bingelis, Tony. 1988. "After That First Test Flight," *Sport Aviation* (April), 27.
_____. 1989. "Flight Testing Homebuilts, Stage Three: Expanding the Flight Envelope," *Sport Aviation* (March), 28.
Hanson, P.W. 1977. "All You Ever Wanted to Know About Flutter But Were Too Smart to Ask," *Soaring* (August), 41.
Meyer, Phillip. 1973. "Flight Testing for Control Surface Flutter," *Sport Aviation* (June), 6.

Owen, Ben. 1986. "Flutter and the Aircraft Craftsman," *Sport Aviation* (November), 66.

_____. 1988. "The Flight Envelope," *Sport Aviation* (March), 42.

Ringer, Bud. 1968. "Designing against Flutter," *Sport Aviation* (August), 28.

Sequoia Aircraft Corporation. 1987. *F.8L Falco Flight Test Guide.* Chapter 3, "Advanced Flight Testing." Obtainable from Sequoia Aircraft Corporation, 2000 Tomlynn St., P.O. Box 6861, Richmond, VA 23230.

Thorp, John. 1976. "Understanding Flutter," *Sport Aviation* (September), 15.

Turner, Eugene. 1969. "The New Look on the Turner T-40, Thorp T-18," *Sport Aviation* (August), 10.

CHAPTER 7

Stability

The subject of stability and control—or handling qualities, as they are often called—is the study of the way an airplane responds to any stimulus, either from the pilot or from the air. As an engineering discipline, it involves determining what response characteristics of the airplane are acceptable and how to improve the ones that aren't. The knowledge of some basic handling-qualities techniques can often help the builder dramatically improve the comfort, and therefore the enjoyment, of the airplane.

The field of stability can be broken into two parts, static stability and dynamic stability. Static stability is defined as the tendency of a system to try to return to, or move away from, a trimmed condition when disturbed. It is normally determined by the direction and force with which the system reacts when it is disturbed from trim. Dynamic stability is defined as the motion with which a system responds when it is disturbed from trim and then set free.

Stability Illustration

Perhaps the easiest way to explain the two types of stability is to use a simple example. Imagine that you have a level tabletop, a large bowl, and a marble. If you put the marble inside the bowl, as shown in Figure 7.1, you have generated a statically stable system. This is because if you move the marble from the center of the bowl, a measurable force will try to

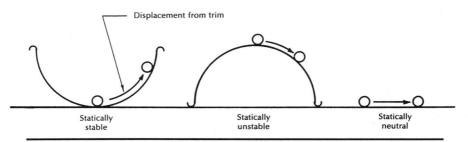

Fig. 7.1. Static stability

move it toward its original position. If you turn the bowl over and balance the marble on the bottom, you will have generated a statically unstable system, because if the marble is moved in any direction from the center, a force is generated that tends to move it away from the center. Taking the marble out of the bowl and putting it on the level table creates a statically neutral system, because the marble tends to stay where it is no matter where that is.

In using the term *stability*, I am referring to the field of study. The terms *stable* and *unstable* describe various levels of stability.

Dynamic stability can also be illustrated with the marble and bowl. If the marble is held against the inside of the bowl and then released, it will oscillate back and forth through the center and up the other side of the bowl. This motion is shown graphically in Figure 7.2. The marble will very gradually lose energy as it rolls back and forth and will eventually come to rest at the bottom of the bowl. This loss of energy is called damping, and the system that behaves this way is referred to as a damped system.

An oscillatory system, as we have with our marble in the bowl, can be completely described, as shown in Figure 7.3, by its period (or frequency, which is $\frac{1}{period}$), its amplitude, and its damping. Damping is defined by the time it takes for the amplitude to decrease to half its original value. This is called the time to half-amplitude, or T½.

If there is no loss of energy, either because there is no damping or because energy is being added to the system at the same rate damping is removing it, you have a system in which the amplitude stays constant.

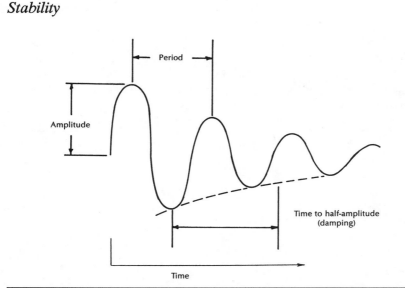

Fig. 7.2. Damped system

Fig. 7.3. Oscillatory system

This undamped or neutrally damped system can be illustrated by gently moving the bowl from side to side in time with the marble so it keeps reaching the same point on the side of the bowl. This is shown in Figure 7.4.

If you move the bowl more rapidly so that the marble climbs higher and higher with each cycle, as shown in Figure 7.5, you have created a divergent, or dynamically unstable, system by adding energy faster than damping can remove it. T½ still applies, but here it describes the time to double amplitude. Normally, divergent systems are undesirable for obvious reasons.

If we greatly increase the damping by adding water to the bowl, the marble will barely reach the bottom of the bowl before it stops, and there will be no oscillatory motion at all. You have now created a critically damped, or deadbeat, system, as shown in Figure 7.6. In aircraft dynamic stability, this is normally the most desirable response.

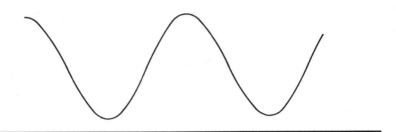

Fig. 7.4. Undamped system

Fig. 7.5. Divergent system

Fig. 7.6. Critically damped, or deadbeat, system

The motion of an aircraft in flight is normally described in terms of angular displacements about the three primary aircraft axes. As shown in Figure 7.7, motions in these axes are normally referred to as the pitch (or longitudinal response), roll (or lateral response), and yaw (or directional response). Because a disturbance in pitch normally does not affect roll or yaw, we will treat longitudinal stability separately. Roll and yaw responses are usually coupled (i.e., they induce responses in each other), so we will treat lateral and directional stability together.

Stability Testing Considerations

Dynamic stability testing is normally done two ways—stick-fixed and stick-free. As you might guess, stick-fixed means with the controls held fixed by your hands and feet, and stick-free means with your hands and feet off the controls. Most conventional control surfaces have a tendency to streamline, or follow the relative wind, somewhat like a flag in a breeze. This tendency, called float, reduces the effectiveness of the control surface as it contributes to stability. Holding the controls fixed with your hands and feet reduces float and increases stability. (An exception to this is a design equipped with a canard. Because the canard is a destabilizing surface, fixing the movable surface on a canard reduces pitch stability. Conversely, releasing the controls in a canard-equipped airplane reduces the effective area of the canard and thus its destabiliz-

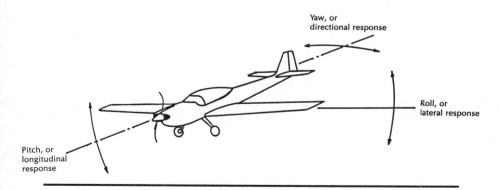

Fig. 7.7. Aircraft axes of angular motion

ing effect, making the airplane more stable stick-free.) The concept of control surface float is important because, as you will see, controlling float is one of your more useful tools in improving stability. We will discuss the concept of float in more detail where it applies to fixing stability problems.

Although the only tools you will need to do stability testing are a stopwatch and a way to keep notes, one piece of built-in test equipment can be very useful. This is a method of measuring longitudinal stick position. Some of the ways this can be done are shown in Figure 7.8. Probably the most sophisticated method is to wire either a potentiometer connected to the stick or a linkage in the longitudinal control system to a meter mounted in the instrument panel. If the base of the stick assembly or part of the longitudinal control linkage is visible, a fixed scale and pointer can also be used. A simpler method is to hang a calibrated strip of metal or wood (like a yardstick) from a pin on the front of the stick and let it slide on a bracket on the underside of the panel. Although this approach can develop significant errors if the stick is displaced to the left or right of center, we will be using it for level flight tests in which there should be little or no lateral stick input. If you use this method, be sure that the strip cannot jam the controls or foul something behind the panel. You should also design this sort of a system so the strip can be broken or easily pulled off the stick if it jams or gets in the way. Another stick measuring system can be designed around a tape measure with the end of the tape attached to the panel. This works well with a control wheel. The tape measure can be attached to the wheel axis so that it remains upright even when the wheel is turned.

All stick indicators should be calibrated against the allowable stick travel. Although inches of stick may be used as a measure, it is probably easier to calibrate your stick position system in terms of percent of full travel. The normal convention is to have full aft be 0 percent and full forward be 100 percent. This system allows you to know at a glance just how much control you have left.

There is one important thing to keep in mind before we discuss how to actually do stability tests. Because you may be looking for subtle effects, be sure to do all such testing in smooth air and with a visible horizon.

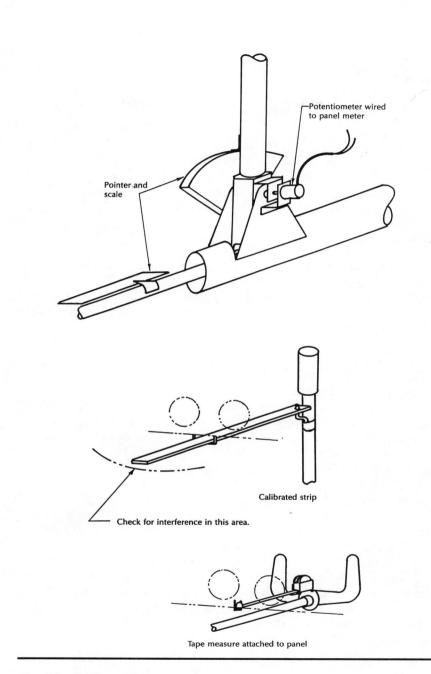

Potentiometer wired
to panel meter

Pointer and
scale

Calibrated strip

Check for interference in this area.

Tape measure attached to panel

Fig. 7.8. Stick position measurement

Longitudinal Static Stability

Static stability is the tendency for an airplane when disturbed from trim to try to return to trim. We cannot measure this tendency directly. However, we can deduce static stability from the control inputs required to hold the airplane at some attitude different from the trim condition. This is roughly equivalent to moving the marble up the side of the bowl and feeling the force required to hold it there. Over the years, a convention has been established as to what constitutes adequate longitudinal static stability.

Static stability is normally measured two ways. The first measures the inherent static stability of the airplane without the effect of the trim system. This is done by determining the relationship between longitudinal stick position and speed, using the control position measuring system described in the previous section.

TESTING WITHOUT TRIM

To do this test, trim the airplane for the slowest speed at which you can comfortably fly hands-off. You should be able to fly hands-off for at least 10 seconds for the point to be considered truly trimmed. Record the stick position and airspeed. Repeat the process every 10 knots up to V_{NE}. Then repeat the points, using decreasing speeds down to your original speed. Do this at forward and then aft c.g.

If you plot the results of this test in the form of longitudinal stick position versus airspeed, the results should look like Figure 7.9. For this curve to be acceptable, the trimmed stick position must move progressively forward as the airplane goes faster. Thus, the plotted stick position must have a positive slope (the curve moves up as the points move to the right) over the whole speed range. The slope will normally be less at aft c.g. and at higher speeds. In fact, a nearly flat slope is acceptable as long as it doesn't become zero or negative. If the slope is unacceptable, you should determine the c.g. at which the slope is acceptable by repeating the test with the c.g. moved slightly farther forward. Keep repeating until the slope is positive at all speeds.

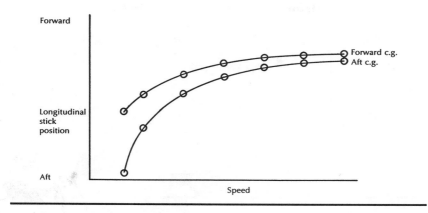

Fig. 7.9.　Longitudinal static stability–stick position versus speed

TESTING WITH TRIM

The second criterion for determining static stability includes the effect of the trim system. Trim to a speed 10 knots higher than the minimum speed you used for the previous test and note the stick position. Without retrimming or adjusting the throttle, slow down 10 knots, let the aircraft stabilize, and note the stick position. Next, push over to a speed 10 knots higher than your trim speed, stabilize, and again note your stick position. Repeat this sequence at several increasingly higher trim speeds, including normal cruise. As before, do this test at forward and aft c.g. If you again plot stick position versus airspeed, the graph should look like that shown in Figure 7.10.

If your airplane uses a spring-centering type of trim system, the new points will be right on top of the line determined by the first set of points (Fig. 7.9). However, if your trim system is aerodynamic (trimmed stabilizer, adjustable tab or servo tab), the results should look like those in Figure 7.10. The slope of a line drawn through each group of points (the points surrounding each target airspeed) should be equal to or more positive than the stick position versus speed curve.

In addition, stick force must change with speed. Slowing down from the trim speed must require an increasing aft stick force, and accelerating to a faster speed must require increasing forward stick-force. Of the three static-stability criteria, that requiring a positive stick-force slope is the most important because a pilot is more sensitive to force than to

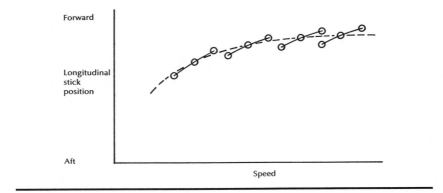

Fig. 7.10. Longitudinal static stability–stick position from trim

stick position. A stick position versus speed slope that is nearly flat is acceptable if the stick position from trim slope is more positive. A shallow stick position from trim slope is okay if you have to apply a noticeable forward force to move to a higher speed and pull with a noticeable aft force to fly at a speed slower than the trim speed.

Longitudinal Dynamic Stability

There are two normal modes of longitudinal dynamic response—one having a long period (low frequency), called phugoid, and one having a shorter period (higher frequency).

PHUGOID

Phugoid is a very slow oscillation in the pitch axis that normally only occurs stick-free. It involves both the basic stability and damping of the aircraft and the characteristics of the trim system. To test for phugoid, trim the aircraft to a reference speed. When the airplane is stable, pull the stick back, without retrimming or changing power, until you have slowed down some incremental speed (10 knots is normal). Once you have stabilized at the new speed, release the stick. The nose will respond by falling through the original trim attitude to some lower attitude. The airspeed will rise until the trim system and the natural stability of the airplane cause the nose to rise again. Normally the nose will overshoot the trim position, and the airplane will again begin to slow down. If the phugoid is damped, the overshoots (cycles) will get smaller and smaller until the oscillation dies out. While the pitch oscillation is going on, you should not touch the stick but use small pedal inputs to hold the wings level.

Record the maximum and minimum airspeed of each oscillation until the oscillation dies out or doubles in amplitude. Use a stopwatch to time a number of complete cycles (minimum airspeed to minimum airspeed) to get an average period. You should test for phugoid at cruise speed and at $1.5V_S$ at both forward and aft c.g. Phugoid damping decreases with decreasing speed, so you can expect the worst characteristics to appear at aft c.g. and $1.5V_S$.

It is acceptable, and in fact common, for the phugoid to be mildly divergent (that is, the amplitude of each cycle is a little larger than the last), providing the period is long. In this case, long means a period of 20 seconds or more, long enough that the pilot corrects the phugoid oscillation instinctively without even realizing that it is an instability. Phugoid can become a problem if it is strongly divergent and/or has a relatively short period (20 seconds or less).

Normally the period of the phugoid oscillation decreases with decreasing airspeed. This fact, combined with the minimum damping normally occurring at $1.5V_S$, indicates that the worst condition for phugoid occurs at low speed and, of course, aft c.g. Thus, the normal approach should be to test forward c.g., then aft c.g. at cruise speed, followed by forward and aft c.g. at $1.5V_S$. A divergent oscillation with a period of 20 seconds or less is unacceptable and should be corrected.

SHORT-PERIOD RESPONSE

The short-period longitudinal response is normally so heavily damped that it is unobservable in airplanes of the speed range of the majority of homebuilts. Unlike phugoid, however, damping of the short-period response decreases with speed, and some of the higher-performance homebuilts may be fast enough for it to become a factor.

The short-period oscillation is excited by using an input called a doublet. This is a rapid forward-and-aft input at about the frequency of the expected response, as shown in Figure 7.11. A doublet is used because it allows you to insert significant inputs at fairly high frequencies without disturbing the basic attitude of the airplane. This is because in a proper doublet the forward and aft inputs are about the same. The inputs should be small to avoid inadvertent overstressing, and for this mode they should be attempted first at a rate of about 1 cycle per second. The short-period oscillation, if it is present, will appear as a relatively rapid nodding, or vertical hunting, of the nose.

If you observe no response, try slightly higher and lower period inputs. First test for the response stick-fixed by holding the stick rigid after inputting the doublet. If the response does not appear, then repeat

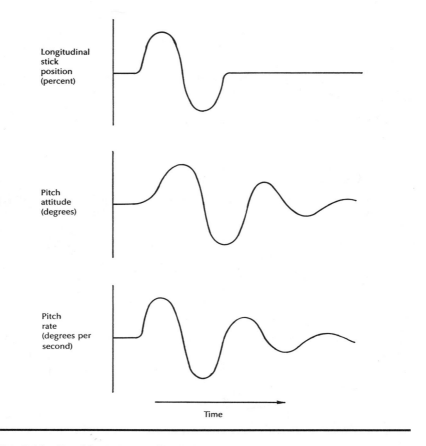

Fig. 7.11. Doublet input and response

the test stick-free by releasing the controls immediately after inserting the doublet.

Because the damping of the short-period response decreases with increasing speed, start testing at low speed ($1.5V_S$). Testing for short-period response should be carried out at speeds up to and including V_{NE}.

Fixing Longitudinal Static Stability Problems

All of the longitudinal phenomena that we have been discussing are affected by the following: the distance between the c.g. and the horizontal tail, the effective area of the stabilizer, the lift-curve slope of the stabilizer, the trim system, and the amount of free play or float in the pitch control system.

C.g. provides the most direct method the pilot has of controlling pitch stability. Moving the c.g. forward increases the effectiveness of the horizontal tail and improves both static and dynamic stability. The primary objective of a stability test program is to prove that the airplane has acceptable stability characteristics at a limiting c.g. This c.g. then becomes the aft c.g. limit called out in the airplane's limitations. If moving the aft c.g. limit forward gives acceptable stability without hurting the utility of the airplane, then this is the simplest fix for a stability problem.

Besides the c.g. location, which affects the distance between the tail and the c.g., the most important factors in determining pitch stability are the area and lift curve slope of the horizontal tail. The lift-curve slope is determined primarily by the aspect ratio of the tail, which is the tail span divided by the average chord. The most powerful way to increase the effectiveness of the horizontal tail is to increase the span, which increases both the area and the aspect ratio. The only fly in this ointment is that increasing the span of the tail dramatically increases the bending loads on the spar and usually requires the design of a whole new horizontal surface. Don't try this option unless you have some capable engineering talent available.

STRAKES

One way of increasing the effective area without increasing the bending loads is through the use of strakes. These are sharp-edged, low aspect-ratio surfaces attached to the fuselage at the root of the horizontal tail, as shown in Figure 7.12. As we shall see, strakes can be used in a number of places to improve stability characteristics in a number of ways. They have several important advantages. Because strakes are sharp-edged, low aspect-ratio surfaces, they do not stall in the conventional sense when flown at a high angle of attack. Instead, a vortex forms along the leading edge and rolls back along the top of the strake. The vortex, because it is a low-pressure area (a little horizontal tornado), not only increases the effectiveness of the strake but also keeps the strake from stalling by mixing high-energy air with the low-energy air on the upper side of the strake. Thus, the strake is more effective than its small area would suggest, and it maintains its effectiveness up to very high angles of attack.

If the strake is located immediately in front of a fixed surface, as shown in Figure 7.12, the vortex will roll back over the fixed surface and

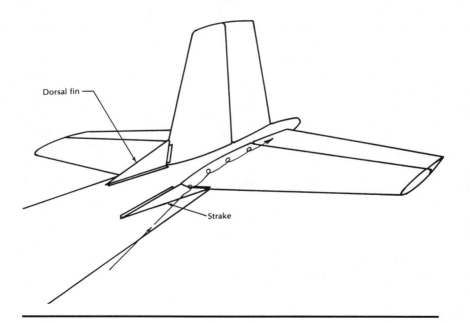

Fig. 7.12. Strakes

also make it more effective and keep it from stalling. Thus, the most effective application of a strake to fix a pitch stability problem is immediately in front of the horizontal stabilizer. This is an enormously useful tool to the homebuilder because it can be added during the test program with no structural analysis required and with little penalty in weight or drag. This is also a common tool for professional engineers. For an example, go look at the tail of a Grumman/American Traveler.

Because strakes can be used in such a wide variety of locations and can be used for a wide variety of purposes, they have acquired specific names for specific applications. For example, when located on the top of a fuselage in front of a vertical fin, a strake is called a dorsal fin. When located under the tail, it's called a ventral fin. Specific manufacturers have also come up with their own names. For example, LERX (Leading Edge Root Extensions) is the name McDonnell-Douglas has given to the strakes located in front of the wings on the AV-8B Harrier.

FLOAT

Another factor that contributes to the effectiveness of the horizontal surface is the degree of float of the elevator or hinged surface. The term *float* refers to the tendency for a control surface to weathervane, or follow the local flow. A simple elevator, hinged at the leading edge and without a trim system, will have almost 100 percent float when stick-free. In this condition it will contribute very little to the stability of the aircraft, thus reducing the effective area of the horizontal tail.

Float can be reduced by restraining the elevator and by reducing the aerodynamic pitching moment of the control surface. Restraining a movable surface in a small airplane is usually done by introducing springs pulling on both sides of the control circuit. The effect would be similar to pressing down simultaneously on both rudder pedals in an airplane with a great deal of rudder float. The effect can be quite dramatic (try it in a Cessna 150). Adding springs not only constrains the control surface but also puts the control system in tension, thus reducing mechanical slop. It also increases control forces. Keep this in mind when we discuss control in Chapter 8.

Because the springs oppose aerodynamic moments, they can be made more effective by reducing the aerodynamic moment. This is usually done by increasing the aerodynamic balance of the control surface. A control surface is typically balanced by adding surface forward of the hinge line. The normal reason for aerodynamic balance is to reduce control forces, so we will discuss ways of increasing aerodynamic balance more thoroughly in Chapter 8.

TRIM SYSTEM

One of the jobs of the trim system is to increase longitudinal static stability. Thus, the longitudinal trim system has a profound effect on stick-force stability and phugoid. Some of the trim systems commonly used are the variable spring, the adjustable tab, the servo tab, and the variable incidence stabilizer. Schematics of some of these are shown in Figures 7.13 and 7.14.

The variable spring, as the name implies, is an adjustable spring or springs hooked to the longitudinal control system. Its job is simply to relieve the pilot's control loads. It can vary in complexity from the simple piano wire and quadrant found in the Duster sailplane to the magnetic brake system found in some helicopters. The spring system will try to hold the controls fixed, and in doing so it will reduce float and thus improve stability. The spring system is simple, light, and creates no drag. When used with aerodynamically balanced surfaces, it can be very effective.

The ground adjustable tab is a simple, sheet-metal extension of the trailing edge of any control surface, as shown in Figure 7.14. It is normally made of annealed aluminum so that it can be bent on the ground to adjust the trim input. It, too, is simple and light. It offers a better response to speed than the spring alone, which will give a better stick-force stability slope. Because the elevator is free to move in the stick-free condition, it may not give adequate protection against phugoid unless the fixed stabilizer is large.

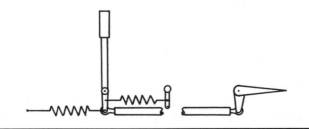

Fig. 7.13. Variable spring

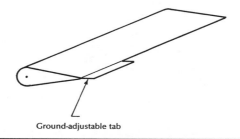

Ground-adjustable tab

Fig. 7.14. Ground-adjustable tab

A simple and very effective trim system can be made by combining a single spring with a fixed tab, as shown in Figure 7.15. The tab should be bent so that increasing speed forces the trailing edge of the elevator up. The spring should be set up to oppose the aerodynamic force on the tab. This system combines the simplicity of the variable spring system with some response to speed.

The servo tab is the most complex and versatile trim system the homebuilder is likely to use. The tab linkage is set up so that the tab motion is coupled with the motion of the elevator, as shown in Figure 7.16. The servo tab is capable of creating very powerful stick gradients with speed or stick position. It is most effective with aerodynamically balanced elevators or stabilators. Servo tabs can allow the use of stabilators that are fully aerodynamically balanced (i.e., hinged at or near the quarter chord). In this case, the servo tab provides all the stabilizing moments and stick forces and results in an aerodynamically efficient installation.

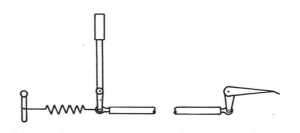

Fig. 7.15. Ground-adjustable tab with spring

Fig. 7.16. Servo tab

Servo tabs are a very powerful aerodynamic device. They are very sensitive to linkage geometry and will usually require considerable tuning of the tab gearing before an acceptable stick-force gradient is attained. For this reason, servo tab linkages should be accessible and should have built-in provisions for geometry changes, as shown in Figure 7.17.

The variable incidence stabilizer is a trim system that was popular for the early Piper products. It consists of a mechanism (usually a cable-driven screw jack) that changes the incidence of the stabilizer. It is simple and effective, and because the screw jack is irreversible, a cable failure simply leaves the trim where it was when the cable failed. The disadvantage of the variable incidence stabilizer is that it doesn't affect the elevator, which is left free to float, thus reducing the stabilizing effect of the whole tail. The screw jack also tends to be a high-maintenance item that requires periodic lubrication and occasional replacement.

If your aircraft has poor pitch stability, one of the most effective fixes is to reduce float. This can be done by introducing a spring- or tab-type trim system, by aerodynamic balancing, and by removing control-system slop. All of these will increase the effectiveness of the horizontal tail.

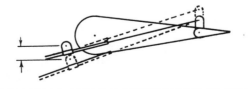

Fig. 7.17. Adjusting the servo tab

Lateral-Directional Static Stability

Independently, lateral (roll) and directional (yaw) static stability are rarely a problem in lightplanes.

In fact, it is acceptable and normal for roll static stability to be neutral or even slightly unstable. What is important is the interaction, called coupling, between roll and yaw responses. This is called dihedral effect, the roll response of the airplane to sideslip. Positive dihedral effect is defined as the tendency to roll in the same direction as a pedal input. Dihedral effect can provide a strong input to help the pilot fly with zero sideslip. An improper amount of dihedral effect may make an airplane difficult to keep in proper yaw trim, and/or it may make a plane pay a drag penalty by being in almost continuous sideslip. An obscure benefit is that dihedral effect allows the pilot, if necessary, to execute turns without the benefit of roll control. Adequate dihedral effect could thus change a jammed aileron from a dangerous situation to one that is merely exciting.

TESTING DIHEDRAL

To test for dihedral effect, first trim for level flight. Slowly and smoothly apply about one-quarter rudder travel. While doing so, use lateral stick to hold a constant heading and longitudinal stick to hold constant indicated airspeed. Smoothly increase the pedal to half, three-quarters, and full travel. Repeat the process in the other direction. At each point observe the amount of pedal force required, the amount and direction of lateral stick deflection and force, longitudinal stick, and the steady roll angle. Dihedral effect tests should be carried no further if you run out of pedal, lateral stick travel, or longitudinal stick travel, or if you experience high pedal forces. Dihedral effect should be checked at several speeds ranging from $1.5V_S$ to V_{NE} within the bounds of the limits given above.

An airplane with positive dihedral stability will require a roll input and roll attitude that are opposite and in proportion to the pedal input. For example, right pedal should require left stick position and force as well as left roll. If the pedal input is doubled, the roll input and attitude should double. Pedal forces may change with sideslip, and in fact the lightening of rudder forces at high sideslip angles is normally acceptable.

A reduction of pedal forces to zero or a reversing of forces, known as rudder lock, is not acceptable. If this were to cause a rudder to go to full travel, it could result in loss of control or structural failure at high speed.

MODIFYING DIHEDRAL

Dihedral effect results from the dihedral angle of the wing, the wing sweep, and the height of the vertical tail. Wing dihedral is the vertical angle of the wing from the horizontal, as viewed from the nose of the airplane. When the airplane experiences a sideslip, the angle of attack of one wing is increased relative to the other, causing the airplane to roll away from the sideslip. Dihedral, or upward tilt, yields positive dihedral effect. Anhedral, or downward tilt, gives a negative dihedral effect.

It is normally difficult or impossible to change the dihedral of the wing. Strut-braced wings can be changed by inserting struts of a different length. The effective dihedral of the wings can be changed slightly by attaching upswept (or downswept) wingtips, as shown in Figure 7.18. Applying dihedral at the wingtips can have other side effects, such as increased induced drag for upswept tips and reduced induced drag for downswept tips. In no case should the new tips be longer than the tips specified by the plans, unless you can verify that the increased root-bending moments will not exceed the structural capability of the wing.

Wing sweep has a significant effect on dihedral, with aft sweep causing increased dihedral effect. Wing sweep is normally introduced by the designer for its other benefits and would often result in excessive dihedral effect if the designer did not take other measures to counteract it. This is why airplanes with large sweep angles often have wings with anhedral. Conversely, if the airplane you are testing has a large wing sweep and no anhedral, you should expect to find problems resulting from excessive dihedral effect.

Almost all airplanes have a vertical tail in which most of the area is above the c.g. When the airplane develops a sideslip, the side load on the tail creates a rolling moment as well as the expected yaw moment, as shown in Figure 7.19. Because the center of area is above the c.g., this rolling moment will be away from the relative wind, which gives the same effect as positive dihedral. You should be aware that any change you make that moves the center of the vertical tail up or down will change the equivalent dihedral. If you increase the height of the tail, effective dihedral will increase. If you add a ventral fin, as shown in Figure 7.20, it will decrease.

Determining what constitutes adequate dihedral effect for your

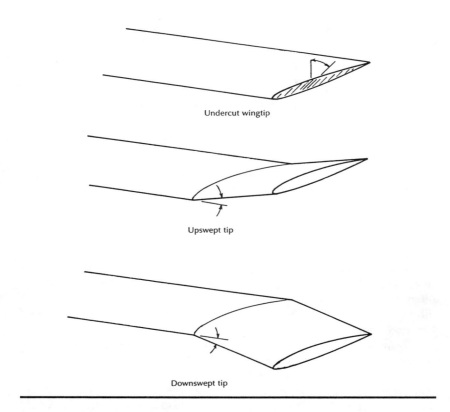

Undercut wingtip

Upswept tip

Downswept tip

Fig. 7.18. Adjusting dihedral

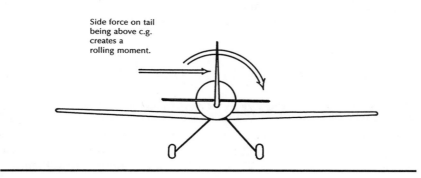

Side force on tail
being above c.g.
creates a
rolling moment.

Fig. 7.19. Effect of tail height on dihedral

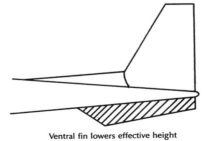

Ventral fin lowers effective height
of tail, reducing rolling moment.

Fig. 7.20. Ventral fin

airplane requires some judgment. There are fewer hard and fast rules for the homebuilder as to the level of dihedral required than there are for pitch static stability. In addition, most of the factors that determine dihedral effect are put in the airplane by the designer and are difficult for the builder to change.

Because dihedral effect is a coupling between two axes of motion, designers usually try to minimize it in airplanes for which pure control response is important, such as aerobatic airplanes. That is why these machines normally have no dihedral, no sweep, and low tails. For an airplane that will be used mostly for cross-country, some degree of dihedral effect is useful to reduce pilot workload in level flight.

At first glance it would seem that very large amounts of dihedral would be desirable for a cross-country airplane. But excessive dihedral can cause increased drag and, more importantly, can work with other effects to create dynamic stability problems, as we will discuss next.

Lateral-Directional Dynamic Stability

There are two modes of lateral-directional dynamic stability that the builder needs to be concerned with. These are Dutch roll and the spiral mode.

DUTCH ROLL

Dutch roll occurs when inadequate yaw stability couples with excessive dihedral effect to create a maneuver in which the nose actually describes an ellipse on the horizon. As shown in Figure 7.21, a sideslip causes the airplane to roll away from the relative wind and slow down. The nose will drop, and the roll will cause a sideslip in the opposite direction. The lower pitch attitude will cause the airplane to accelerate and pitch up while the roll causes a sideslip in the original direction, and then the process begins again. Dutch roll usually occurs in airplanes with a large dihedral effect and poor yaw stability. It is common in swept-

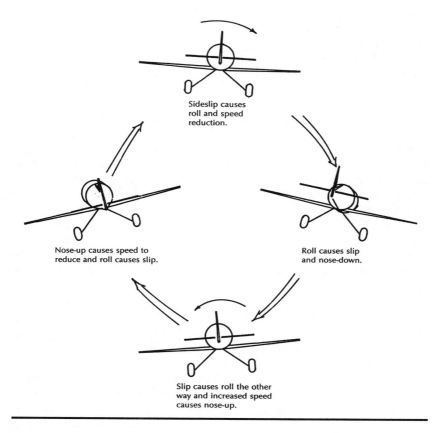

Fig. 7.21. Dutch roll

wing business jets and transports and is usually controlled in these airplanes by a yaw damper that senses yaw rate and introduces rudder inputs to damp the yaw response.

To test for Dutch roll, trim for level flight at the selected speed. Then introduce a rudder doublet just as you introduced the longitudinal doublet when looking for the short-period pitch response. Put in small left- and right-pedal inputs of the same size and try to match the frequency of aircraft response you feel. Dutch roll should be checked both stick-free and stick-fixed, with the difference being the effect of float. This is particularly important in Dutch roll, when the control of float is an important tool in improving poor characteristics.

Record the number of cycles until the oscillation dies out, and record the frequency of the oscillation. If there are 6 or more cycles but the oscillation does die out, the response is considered lightly damped. Although this is sometimes considered acceptable, it is normally uncomfortable in turbulence or following abrupt control inputs and should probably be improved. If there are fewer than 6 cycles in the hands-off mode, the response is good and does not require any improvement.

The frequency of the Dutch roll gives an indication of yaw stability. A frequency of less than 1 cycle per second (period greater than 1 second) is an indication of poor yaw (directional) stability.

The final parameter needed to describe Dutch roll is the ratio of roll to yaw during the oscillation. Roll-yaw ratio is easiest to determine by observing the motion of a wingtip against the horizon, as shown in Figure 7.22. If the tip describes a circle, the airplane is yawing as much as it is rolling and the roll-yaw ratio equals 1. If the tip describes an ellipse, the ratio is less than 1 if the ellipse is horizontal and greater than 1 if the ellipse is vertical. Ratios greater than 1 are undesirable and normally require a great deal of damping to give comfortable characteristics.

As you can see, Dutch roll is a complex oscillation involving several interlocking relationships. In general, the stick-free Dutch roll should damp out in 6 cycles or less, have a frequency of 1 cycle per second or slightly greater, and have a roll-yaw ratio of less than 1. As I pointed out, Dutch roll usually occurs in airplanes with excessive dihedral effect and inadequate yaw stability. You would think that reducing dihedral and increasing yaw stability would improve Dutch roll. Unfortunately, it's not that simple. Increasing directional stability will increase the frequency of the oscillation without significantly affecting damping. If stability is increased by increasing the height of the vertical tail, which

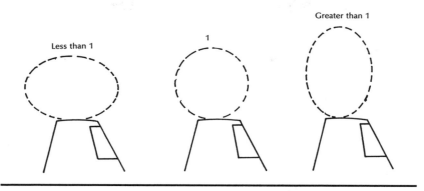

Fig. 7.22. Measuring roll-yaw ratio

increases dihedral effect, it will make the roll-yaw ratio worse. Decreasing dihedral increases damping, decreases frequency, and reduces roll-yaw ratio. Damping is also increased by reducing rudder float.

It is normally desirable to reduce dihedral effect, increase yaw stability, and increase damping. One good way to do this is with a ventral fin. This fin increases yaw stability while reducing the tail contribution to dihedral effect. In addition, reducing rudder float will improve damping. Rudder float can be reduced by adding rudder-centering springs and aerodynamically balancing the rudder.

Another way of increasing yaw stability without increasing dihedral effect is through the use of a dorsal fin, as illustrated in Figure 7.12. The dorsal fin has the advantage of creating a significant improvement in yaw stability without requiring structural changes to the tail. Although a dorsal fin does add tail area, this is not its primary benefit. As a low aspect-ratio body, it is an inefficient lifting surface. It does, however, generate a large vortex at even small sideslip angles. This vortex flows down the side of the vertical fin away from the relative wind, creating a low-pressure region and a fairly powerful stabilizing force.

When testing for Dutch roll, you may get a pure yaw oscillation instead of the classic roll-yaw coupling. In this case the wingtip will move forward and aft on the horizon with little or no vertical motion. This is called snaking, and it is caused by inadequate yaw stability in an airplane with little or no dihedral effect. The same general rules apply to snaking that apply to Dutch roll: it must damp out in less than 6 cycles and have a frequency greater than 2 cycles per second. The yaw-mode fixes for

Dutch roll will also work for snaking: increase the tail power with a dorsal fin, ventral fin, or additional area, and reduce rudder float to increase damping.

SPIRAL MODE

In comparison with the complexities of Dutch roll, the spiral mode is quite simple. Spiral stability is defined as the tendency for an airplane to change its angle of bank while in a turn. The tendency to reduce bank is called spiral convergence and the tendency to increase the bank angle is called spiral divergence.

Test for spiral stability by trimming for level flight and then rolling into a shallow turn (15 to 20 degrees). Once in a steady-state turn, release the controls and measure the time it takes for the airplane to roll to half the initial bank angle if convergent, or to twice the angle if divergent. Repeat this at several different speeds and at both forward and aft c.g.

Spiral divergence is a relatively subtle effect. If the aileron system friction does not allow the ailerons to return to neutral, spiral divergence can be affected. For this reason you should try to return the stick to the center of the friction band before releasing it. You should also repeat the test several times in both right and left banks, looking for consistent results.

In general, spiral stability should range from moderately stable to slightly unstable. There is no real improvement in handling if the spiral mode is strongly convergent. However, a divergent time to double amplitude of 20 seconds or less is strongly unstable and unacceptable. The reason that a neutral or even slightly unstable spiral mode is okay is that the increasing bank angle develops so slowly that the pilot subconsciously controls it with small aileron inputs.

Like Dutch roll, spiral stability is a result of a combination of dihedral effect and yaw stability. In this case, it is inadequate dihedral effect coupled with a strong yaw stability. Because it is not normally good practice to reduce yaw stability, the best way to fix excessive spiral divergence is to increase dihedral effect.

General Stability Considerations

As you can see, stability testing is a complex game demanding knowledge, skill, and patience. If done properly it can increase your skill and help you develop a nice-flying airplane.

Stability effects are often subtle and easily masked by turbulence. For this reason, stability testing should always be done in smooth air, even if this means early morning or late evening flights. Because observation of accurate attitude is important, it should also be done only when there is a distinct horizon visible.

You may find it difficult to get consistent results. Have patience and try the test condition several times, experimenting with input size and frequency until a pattern emerges.

In general, lightplanes are most stable at forward c.g., stick-fixed, and at moderately low speeds. Start from here and build up to aft c.g., high speed, and the stick-free condition. In all cases, use small inputs. Large ones tend to mask true effects or make things happen too fast for you to see the true effect.

Further Reading

Collinge, George. 1984a. "Is a Horizontal Tail Necessary," *Sport Aviation* (April), 27.

_____. 1984b. "Pitch Stability Continued," *Sport Aviation* (April), 27.

_____. 1984c. "Lifting Tails, One Piece Tails and Servos," *Sport Aviation* (June), 38.

_____. 1984d. "Pitch Stability Further Examined, Pt. 4," *Sport Aviation* (July), 39.

_____. 1984e. "Tailless, Pt. 5," *Sport Aviation* (August), 38.

_____. 1984f. "Pitch and Roll Retrospect," *Sport Aviation* (September), 24.

Erickson, James. 1975. "Aircraft Dynamics—What Pilots Never Ask," *AIAA Student Journal* (Spring), 10.

Garrison, Peter. 1989. "Balancing Act," *Flying* (May), 90.

Schiff, Barry. "Yaw Damper for Bonanzas," *AOPA Pilot* (February), 62.

CHAPTER 8

Control

If stability is defined as the way an airplane responds to any stimulus, control is the way an airplane reacts to a stimulus from the pilot. Control is normally defined by two parameters—response and feedback. Response is the static and/or dynamic reaction to a given control input by the pilot. Feedback is the way the airplane tells the pilot, through control force and position, what it is doing.

Adequacy of control, like stability, is judged against a series of criteria. Unfortunately, many of these criteria require special instrumentation, putting their measurement beyond the capability of the average homebuilder. Fortunately, exact adherence to these criteria is not required to have a safe, flyable airplane. We can do a perfectly adequate job using qualitative and approximate methods.

One of the biggest factors that determines control characteristics is the basic stability of the airplane. In addition, we have been using control response to determine stability. Thus, many of the tests we just used to define stability can also be used to answer important questions about control.

Longitudinal Control

The two most important criteria for longitudinal or pitch response are the ability to achieve the desired speed range in all normal configurations and the ability to achieve the desired load factor. We actually started developing data on pitch response when we tested for longitudinal static stability by looking at longitudinal stick position when compared with speed. We should use the same type of curve, but this time we are looking not at the slope of the line but at whether the pilot can achieve the full speed range with some control remaining. Figure 8.1 shows a plotted example of a nearly ideal longitudinal stick position versus speed. The pilot has at least 10 percent forward stick remaining at V_{NE} and the worst c.g. There is also stick remaining at V_S and forward c.g.

We also need to look at the effect on control of configuration changes made by landing gear and flaps. You will note in Figure 8.1 that the only time the pilot requires full stick authority is at minimum speed in the landing configuration (with flaps and gear down). This is the only time that having less than 10 percent control margin is both acceptable and normal.

The ability to attain the full load-factor range is also something you have already determined. When you did the envelope-expansion flights, you flew to the limit load factors with available control. If the load factors you achieved were limited by control response, they then became the limit load factors for your airplane. If these load factors are unsatisfactory, then you will need to increase the pitch-response

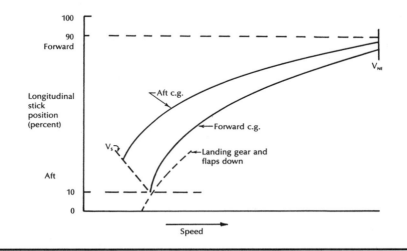

Fig. 8.1. Longitudinal control

capability of your control system.

As a pilot, you determine how an airplane is responding to control inputs by using your eyes and middle ear, which tell you what the aircraft's attitude is and how that attitude is changing. In order to do this, however, you must wait for the aircraft's attitude to begin to change. A more immediate indication of what the aircraft is going to do is the motion and force that you apply to the controls. This motion and force is normally referred to as feedback. It is the way the aircraft signals the pilot how it is about to respond. It turns out that the pilot is not very sensitive to small changes in control position but is very sensitive to small applications of control force. For this reason, although we check stick-position feedback, good stick-force feedback is more important.

DETERMINING STICK FORCES

Longitudinal-force feedback needs to satisfy two criteria: stick force with speed and stick force per g. Stick force with speed is something we already determined when we tested for stick force from trim when doing longitudinal static stability. As we pointed out at that time, the criterion is that reducing speed from trim requires aft stick force, and increasing speed requires forward stick force. If we had the instrumentation installed in the airplane to measure the force and plot the results, the curve should look like that in Figure 8.2.

Stick force per g is the amount of aft stick force you must apply to increase load factor, as you would in a turn or pull-up. At first glance you would think that we would design the control system to achieve as

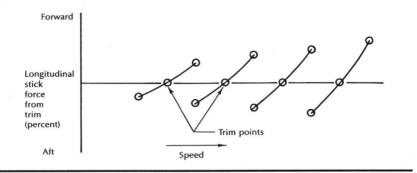

Fig. 8.2. Longitudinal stick force with speed

low a stick force per g as possible. This is not desirable because stick force is one of the most important cues a pilot uses to determine what the airplane is doing. A very low stick force per g can make precise pitch control difficult, increase the chance of pilot-induced oscillation, and, most importantly, increase the chance of pulling too many g's and damaging the airplane. It is very likely that the extremely low stick forces found in some ultralight aircraft have been a factor in some in-flight structural failures, simply because the pilot overcontrolled.

Determining stick force per g demands more complex instrumentation than most homebuilders have access to. The easiest way for you to do it is to select a maneuver point and tune the control system until the stick force at this point feels the way you want it to. A good point to use for a nonaerobatic airplane is a 60-degree bank turn in level flight, which will yield 2 g's. This maneuver should require a firm pull on the stick but should be achievable with one hand. For an aerobatic airplane, lighter control forces are desirable, so a more severe maneuver, such as a 3- or 4-g symmetrical pull-up, should be used. It's obvious that the stick force per g resulting from this type of qualitative testing will yield results that differ from pilot to pilot. The differences are not critical, and it is more important for the pilot to pull with an appreciable force to develop load factor than to match an exact load. This is also your airplane, and you will tune the airplane to what feels good to you. In any case, it is probably better to err on the side of higher stick forces rather than lower.

ADJUSTING STICK FORCES

Longitudinal stick forces are determined by the amount of aerodynamic balance of the elevator, the control system gearing, the design of the trim system, and the presence and design of an elevator bobweight.

Aerodynamic balance is proportional to the distance between the aerodynamic center of the elevator and its hinge line. As we discussed in Chapter 7, moving the aerodynamic center forward, closer to the hinge line, reduces the control surface moments and therefore the stick forces. Conversely, if the distance between the aerodynamic center and the hinge line is large, as it would be for an elevator hinged at its leading edge, the longitudinal stick forces will be large. If this distance is small, as it would be for a stabilator hinged at its quarter chord, the stick forces will be small and may even approach zero. For elevators there are two ways to increase the aerodynamic balance, as shown in Figure 8.3—the displaced hinge and the aerodynamic counterbalance. The displaced hinge simply moves the hinge line aft, closer to the center of pressure. Although an elegant solution, it must be in the initial design and is very difficult to change once the airplane is built. The aerodynamic counterbalance is not quite so pretty, but it is easier to modify to achieve the desired pitch forces.

The gearing in the pitch control system can sometimes be used to adjust stick forces if they are a result of excessive or inadequate control sensitivity. Be careful, however, because altering gearing also changes the design loads in the control system and the available elevator travel. Be sure that you will not generate excessive control loads and that you have

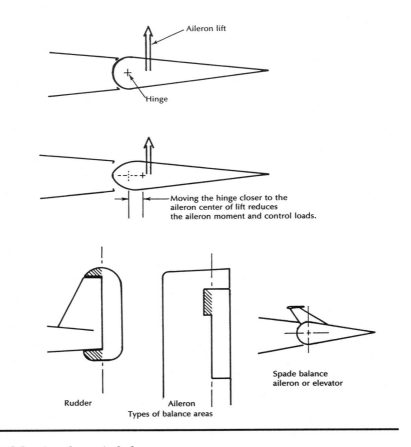

Fig. 8.3. Aerodynamic balance

adequate elevator travel to fly safely. You may have to make some new control stops to accommodate the new control travel.

A trim system supplies stick force per g automatically because it resists the control deflections and changes in control loads that happen in any accelerated maneuver. Whether you have a spring-centering, aerodynamic, or combination system, it will compel you to apply some force to achieve a desired load factor. In many cases, however, this does not give sufficient stick force per g. Moreover, the things that determine the relationship between stick force and load factor, such as aerodynamic balance, trim system characteristics, and control gearing, are determined by other factors (and are a pain in the neck to change anyway). In this case, the designer can use a device called an elevator bobweight as a tool to get a good stick force per g.

A bobweight is simply a mass built into the control system that tends to pull the stick forward when the load factor increases. One such system is shown in Figure 8.4. Normally the mass is installed as close to the stick as possible to maximize the force at the stick. Do not locate the mass at or near the elevator (or stabilator) because this could affect the flutter characteristics of the control surface.

You will notice that Figure 8.4 shows the bobweight in conjunction with a set of tabs and a spring. This allows each system to make up for the deficiencies of the others. The spring supplies stick-force gradient at low speeds when the tabs are less effective and puts the control system in tension, thereby reducing float. The tabs counterbalance the spring

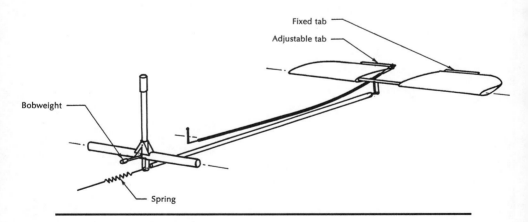

Fig. 8.4. Combination trim system

and help provide stick-force gradient with speed for a balanced surface. The servo tab provides variable trim and an adjustable stick-force gradient. The bobweight provides whatever additional stick force per g is required for good handling during accelerated maneuvers.

Lateral Control

The lateral (roll) control system is normally designed to develop a given roll acceleration and rate from a specific input. You can test for this by using what's known as a step input. An ideal step input and the resulting angular rate and acceleration are shown in Figure 8.5. Unfortunately, developing this type of data requires instrumentation well beyond the capability of most homebuilders, so we will again fall back on our qualitative methods.

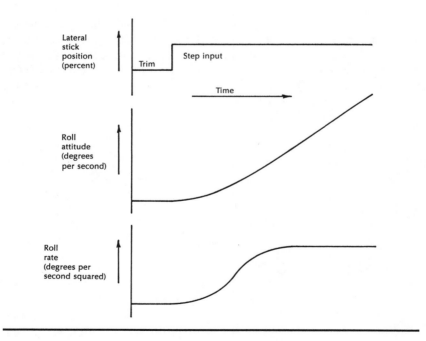

Fig. 8.5. Roll response

To test for dynamic roll response, trim the aircraft in a level turn at a moderate angle (say 45 degrees). Fix the pedals. Then put in and hold a small step input away from (to the outside) of the turn. Hold the input as long as is comfortable (at least to a 45-degree bank in the other direction). Try inputs in both directions at about ¼, ½, ¾, and full travel if possible. At a minimum, this should be done at $1.3V_S$ (approach speed) and V_C (design cruising speed).

Each step input should result in a smooth acceleration to a steady rate of roll with no hesitation or severe yawing. Increasing the input should result in a proportional increase in acceleration and roll rate. Half-stick input at approach speed should result in a significant roll rate, and roll-stick forces should not be so high that full lateral travel is unattainable at V_C.

AILERON EFFECTIVENESS

Aileron effectiveness is determined by aileron area, control deflection, leakage, separation, wing torsional stiffness, and required stick force (feedback). The contributions of aileron area and deflection are fairly obvious. If aileron response seems poor at low speeds, the first thing to check is that the control rigging allows the aileron travel called for by the plans. Modifying the rigging to allow more deflection may help. However, aileron effectiveness starts to deteriorate at angles greater than 15 degrees, and at large angles you risk stalling the aileron. For this reason, angles of greater than 30 degrees are not helpful with normal ailerons.

Aileron area is normally determined by the designer and is an element over which the builder has little control. Most homebuilt designers make their ailerons larger than required, which tends to lead to high lateral-control forces unless the ailerons are aerodynamically balanced. As a crude rule of thumb, the aileron area should be about 10 percent of the wing area. Excessive aileron size results in large roll-control forces and the use of only small roll inputs. In this case, reducing the aileron gearing so that the same stick travel will result in less aileron travel may help. Be careful not to reduce your aileron travel too much, or not enough control will be available at low speeds.

You can also increase the effectiveness of the ailerons (or any other control surface) by reducing leakage around the leading edge of the control surface. As shown in Figure 8.6, this can be as simple as putting tape over the gap for a control with hinges located on the top or bottom of the wing, or adding a foam strip or brush on the radius. More complex internal sealing devices can be used if the aileron uses internal leading-edge balances, as shown.

The disadvantage of seals is that they tend to require maintenance. Both tape and plastics degrade with time and environment, and they could break and leak or come loose and get into the aileron. They need to be inspected periodically, and tape seals should be replaced annually at a minimum. Wipers also need to be installed carefully to minimize gaps between the surface and the seal. Wipers also cause drag by scrubbing on the control radius. This increases control loads and tends to prevent the ailerons from recentering if the control surface moments are small. This can be a problem because it makes it difficult for the pilot to precisely find the control center by feel.

Aileron effectiveness can also be affected by inadequate torsional rigidity of the wing, which allows the wing to twist in opposition to an

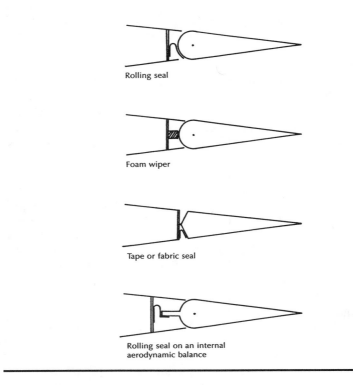

Rolling seal

Foam wiper

Tape or fabric seal

Rolling seal on an internal aerodynamic balance

Fig. 8.6. Gap seals

aileron input. What happens, as shown in Figure 8.7, is that the down aileron tries to twist the wing's leading edge down. If the wing is not stiff enough in torsion, the wing will actually twist, causing the wingtip to act like a huge tip aileron. Because it has a greater area than the aileron, it can overpower the aileron roll input. This can cause a reduction in roll rate, especially at high speeds, and in the extreme case it can cause roll reversal. Inadequate wing torsional stiffness should be suspected if large- or full-authority stick deflections can be achieved in the cockpit but roll rate does not increase (or actually decreases) with increasing speed. This condition most often arises when the aircraft has long, flexible wings, like those found in gliders.

Unfortunately, fixing this condition requires that the wing be stiffened in torsion, which is not easy. It can be done by increasing the thickness of the wing skin, increasing the thickness of the spar webs of a box spar, or by replacing a single strut with a V-strut.

Maximum roll rate is proportional to V^2. Unfortunately, the lateral stick force for a given deflection also increases with speed. As shown in Figure 8.8, maximum roll rate at high speeds may not be achievable because the lateral stick force becomes too high for the pilot to achieve full stick deflection. The roll axis is more sensitive to this than the pitch or yaw axes because the human arm cannot apply as much force sideways as it can fore and aft. This is especially true for side-stick controllers because the side force comes primarily from the wrist and forearm.

Roll stick force is normally reduced by increasing aerodynamic balance. This is done with a displaced hinge or aerodynamic balance area in the same way it is done for the elevators (Fig. 8.3).

Another way of increasing the aerodynamic balance for the ailerons is through the use of Frise ailerons. Because these are normally used as tools to reduce adverse yaw, I will discuss them in a later section.

You can also provide both increased aerodynamic balance and sealing by using an internal seal balance, as shown in the last drawing of Figure 8.6. In this case, the balance area of the aileron is connected to

Fig. 8.7. Aileron reversal

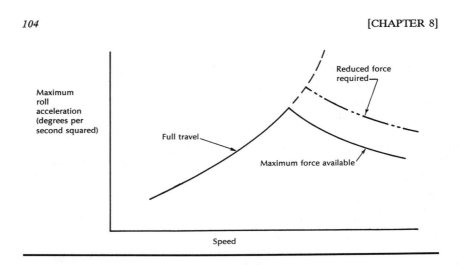

Maximum roll acceleration (degrees per second squared)

Reduced force required

Full travel

Maximum force available

Speed

Fig. 8.8. Maximum roll control

the rear spar with a flexible seal. The balance area is covered by an extension of the wing skin that approaches the aileron. Air pressure vented through the gap between the aileron and skin presses on the seal to relieve aileron loads. This type of balance is very effective and is not sensitive to rigging. Because it has a flexible shroud buried inside the wing, it may require more maintenance than a Frise aileron.

AILERON-ADJUSTMENT DEVICES

Several types of trailing-edge devices can also be used to reduce aileron loads. Of these, two—fixed tabs and geared tabs—could be useful to the homebuilder. The geared tab is similar to the geared or servo tab described in Chapter 7, and it functions in the same manner. Although it is complex, adds drag, and reduces aileron effectiveness, it has the advantage that, if properly designed, it can be adjusted to give just the right amount of roll stick force.

Fixed tabs, again similar to the fixed tabs on the elevators, can be used on ailerons provided that they have sufficient differential gearing. In this case, both the tabs should be bent down to cause aileron upfloat. Because the aileron that is moving up is more highly geared (and has a higher mechanical advantage), its tab will be more effective and will reduce the lateral stick force at large deflections. This system is simple and easy to tune, but is not effective for small deflections or with aileron systems with little differential gearing. Even if roll stick force is not anticipated as a problem, it is very smart to add ground-adjustable tabs to the ailerons in the construction phase because they can also be used to alleviate wing heaviness.

Sometimes aileron forces are too light or require very little force to get large deflections. This can lead to difficulty in precise roll control and can cause pilot-induced oscillation or structural damage at high speeds due to overcontrol. Lateral stick force can be increased by the opposite of the measures described above (altered gearing, tabs, removal of balance area, etc.). Stick force can also be increased quickly and easily with the addition of centering springs. Springs are simple to install, nearly foolproof, and easy to adjust. They can be installed at the stick if necessary.

Springs have two disadvantages. First, because springs do not respond to air pressure, the stick force for a given deflection does not change with speed. This gives the aileron (or any other surface using only springs to supply control force) an artificial feel. It also increases the possibility of overcontrol at high speed because the stick force does not increase with speed. The second problem is that springs are constantly flexing and thus are prone to wear, fatigue, and corrosion. They should be inspected periodically and replaced at a fixed interval, say every two years.

Ailerons can also lose effectiveness because of local separation or stall (Fig. 8.9). The local separation can occur on the aileron from excessive travel of the control surface or on the wing forward of the aileron. It will appear as a dramatic or complete loss of roll control at low speed, possibly accompanied by yawing. This is a potentially dangerous problem because it could result in a loss of roll control or a violent sideslip just at the worst time—on landing. Local separation can

Aileron separation causes loss of effectiveness
at low speeds or large deflections.

Fig. 8.9. Aileron separation

be determined by tuft testing, as described in Chapter 10.

Once the area of separation has been defined by the tuft test, there are several methods that can alleviate local separation. Two that are practical for the homebuilder are slats or vortex generators. Both of these function by taking high-energy air from beyond the boundary layer and mixing it with the separated low-energy air.

Slats and slots, as shown in Figure 8.10, take high-energy air from the leading edge or underside of the wing and squirt it into the low-energy air on the top of the wing to keep it attached. The wings of the Temco Swift and the Stinson 108 are good examples of the use of slats. Slots and slats are very effective, but they are difficult to build and have a drag penalty. The biggest hindrance to the use of slots and slats by the homebuilder is that they must be designed and built in to the wing and are thus extremely difficult to add to an existing wing. A word to the wise though: if any airplane, production or homebuilt, has slots or slats, they were put there for a good reason. Altering or removing them is a quick way to give an airplane really scary low-speed characteristics, such as loss of roll control, wing drop, or poor stall/spin characteristics.

A more useful way of controlling local separation after the wing is built is through the use of vortex generators. As shown in Figure 8.11, these look like small teeth sticking up at an angle to the airflow. Each tooth forms a vortex that trails behind it like a small tornado, mixing the low-energy boundary layer with the high-energy air above it to reattach the flow. Vortex generators may appear crude, but they are simple, effective, and flexible, and they are thus a commonly used item in the aerodynamicist's bag of tricks. They are found, among other places, on the wings of the Boeing 707 and the early Learjets. They are also now available as retrofit items to improve performance on some twins (McClellan 1989).

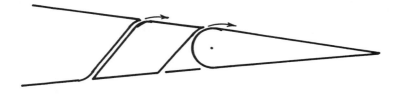

Fig. 8.10. Slots or slats

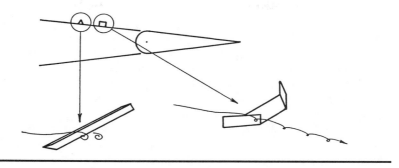

Fig. 8.11. Trip strips, or vortex generators

ADVERSE YAW

One type of roll response that is not desirable and that you are usually trying to minimize is a phenomenon known as adverse yaw. Adverse yaw is a yaw or sideslip away from the direction of roll. It is caused by the different drags developed by the up and down ailerons.

Let's suppose we have a wing with a simple aileron hinged at the leading edge at each end. If one aileron is rotated down and the other up, they will generate up and down forces respectively, which will cause the airplane to roll. If you remember basic aerodynamics, you'll remember the form of drag, called induced drag, that is created as lift is created. Thus, the down-moving aileron, which is creating more lift, will have more drag and will tend to lag, causing the airplane to yaw in the opposite direction to the roll input. This tendency to yaw away from the roll input is called adverse yaw.

Older airplanes, some gliders, and many homebuilts have large amounts of adverse yaw and require large pedal inputs to get a coordinated turn. These are often known as rudder airplanes.

To test for adverse yaw, trim for level flight, release the pedals, apply a lateral step input, and hold it. Watch carefully for the initial direction of motion of the nose as the input is applied. As the roll angle becomes appreciable, the airplane will develop a sideslip, and its natural directional stability will cause it to yaw into the roll, hiding the adverse yaw.

If the nose moves laterally away from the direction of roll (left in a right roll), you have adverse yaw. If the nose slides into the roll, you have proverse yaw. Most designers try to make adverse or proverse yaw as small as possible.

There are two common ways of controlling adverse yaw: differential travel and Frise (pronounced *freeze*) ailerons. Displacement of an aileron into the slipstream creates profile drag whether it is creating or reducing lift. Thus, if we displace the aileron that is moving up more than the one that is moving down, it will create more drag to oppose the increased induced drag of the down aileron. History has shown that a ratio of about 2 to 1 nearly eliminates adverse yaw for significant aileron deflections. If you set up your control system so that the up-moving aileron deflects twice as far as the down-moving aileron for large deflections, you will probably be disappointed to find the adverse yaw still present for small roll inputs. This is because the control-system parts move in a series of radii, giving a very nonlinear response. It takes an appreciable amount of deflection for the differential motion to appear, as shown in Figure 8.12.

A simple way to alleviate the problem of adverse yaw for small deflections is to rig the controls so that the ailerons are a couple of degrees up when the stick is in the neutral position. Small stick deflections will initially reduce the deflection of the down-moving aileron while the up-moving aileron position is continually increasing. By the time the down-moving aileron gets to zero, the up-moving aileron will have an appreciable deflection and you will have a built-in differential. As the control-surface deflections become larger and the up-displacement you rigged in at the neutral stick position becomes insignificant, the

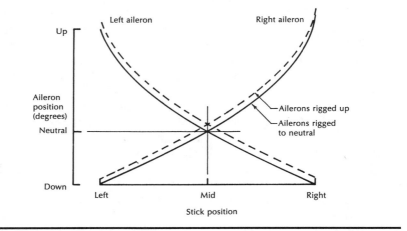

Fig. 8.12. Differential aileron rigging

differential gearing built in to the control system will become effective.

Another way to solve the adverse-yaw problem is with Frise ailerons, as shown in Figure 8.13. The Frise aileron functions as a normal hinged aileron when deflected down, but when it is deflected up, a sharp-edged spoiler is exposed below the wing. The spoiler creates additional drag to oppose the increased induced drag of the down aileron. The Frise aileron has a nice side effect in that the spoiler serves as an aerodynamic balance to reduce lateral stick forces.

Although Frise ailerons help solve both balance and adverse-yaw problems, their proper performance is sensitive to both their design (shape) and rigging. If improperly rigged, the Frise aileron can experience separation on the lower surface, which can cause aileron shudder, snatch (sharp changes in stick force), and loss of authority. Another problem with Frise ailerons is their tendency to float up at high speeds because of increased airloads and cable stretch. This can cause an unstable aileron system that will be very difficult to center. These problems can be reduced by raising the nose shape of the aileron, as shown in Figure 8.13.

A third way to reduce adverse yaw is by the use of spoilers instead of ailerons for roll control. Roll spoilers, as shown in Figure 8.14, are normally located at about 60 to 70 percent of the chord and are

Normal Frise aileron

Frise aileron with raised nose shape

Fig. 8.13. Frise ailerons

themselves 3 inches or less in chord. Spoilers do give excellent roll control with little or no adverse yaw, but like most things they have their deficiencies. Unlike ailerons, spoilers are not under load when at small deflection, so they do not provide adequate aerodynamic feedback. In fact, spoilers will float up into the low-pressure region over the wing, so they are normally pulled closed by springs. As a result, lateral stick feedback is provided by springs and there is little or no aerodynamic stick force, giving an artificial feel to the controls. Some aircraft, like the World War II P-61 Black Widow and the Aerospatiale TBM 700, use small, unbalanced ailerons with their spoilers to supply adequate aerodynamic feedback.

Even with small ailerons providing aerodynamic feedback, lateral stick forces tend to be low with spoilers, so it is wise to use return springs strong enough to reduce the chance of overcontrolling in roll at high speeds. Because the spoilers have to penetrate a rather thick boundary layer before they cause a rolling moment, they often have a so-called dead band of ineffectiveness at mid travel. This is more of an annoyance than a problem. It can often be reduced by rigging the spoilers to be slightly up when the stick is at mid position. This should only be done if the dead band is present throughout the speed range. Otherwise you risk affecting performance.

The control response of spoilers is related largely to their chord position on the wing. When located forward of 60-percent chord, they are very powerful. However, they develop a time lag between input and response that increases as they are moved forward. When moved progressively aft from the 70-percent chord position, their effectiveness decreases and they develop an increasingly large dead band.

Because spoilers affect lift, they suffer a dramatic loss of control power at zero or negative g's. For this reason they are not used when control at low load factor is important, as in aerobatic airplanes.

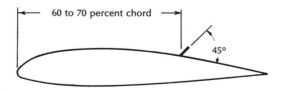

60 to 70 percent chord

45°

Fig. 8.14. Spoiler

Spoilers seem to work best in airplanes of fairly high wing loading. Roll spoilers have been tried several times that I know of in sailplanes and have not worked well. It may have something to do with the small spoiler having difficulty penetrating the thick boundary layer found at these low Reynolds numbers. An additional impediment to using spoilers on sailplanes is that during the takeoff roll, when roll control is critical, wing lift is low, and thus roll control would be poor.

A variation on the spoiler theme, called the slot-lip spoiler, can be usefully applied to an aircraft equipped with full-span fowler flaps. As shown in Figure 8.15, the slot-lip spoiler functions as a normal spoiler when the flaps are retracted. When the flaps are extended, it forms the trailing edge of the wing in front of the flap. In this position roll control is actually increased when flaps are extended, giving improved control at low speeds while allowing the use of full-span flaps.

Directional Control

Yaw, or directional, control is different from either pitch or roll control in that it normally is used to simply maintain zero sideslip (to keep the fuselage aligned with the relative wind) by countering the effects of fuselage instability, turbulence, torque, or adverse yaw. The rudder is used as a primary attitude control only to assist the ailerons in airplanes with poor roll control or during sideslips. For the purposes of this discussion, we are going to ignore the case of a twin with one engine failed.

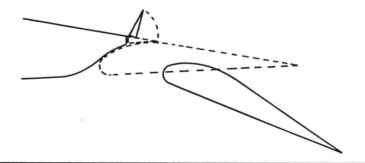

Fig. 8.15. Slot-lip spoiler

For single-engine airplanes, yaw-control power is normally determined by takeoff or landing requirements or by the ability to control the aircraft in yaw when the landing gear is on the ground. Feedback does not have to be tuned as carefully as in the pitch or roll systems since the pilot's feet are less sensitive than the hands and are capable of applying much more force. In fact, as long as yaw stability and control during takeoff and landing are adequate, rudders can be quite small in proportion to the total tail area, as in the Wittman Tailwind.

DETERMINING YAW CONTROL

Yaw control is determined by using the same constant-heading sideslip that you used to measure dihedral effect. Trim for level flight. Then apply about ¼, ½, ¾, and full pedal (if possible) while holding the heading with aileron and the speed with longitudinal control. At each pedal position, note the roll angle and pedal force required to hold the desired heading. Repeat the exercise in the other direction. Roll angle and pedal force should increase with increasing amount of pedal input. The pedal force should increase enough so that you must apply a perceptible increase in force at any position to get a distinct change in sideslip angle. If the force required to move the rudder is too small, you will find it difficult to hold a given sideslip angle accurately.

The pedal force required can be increased by adding pedal springs or by reducing the aerodynamic balance. The aerodynamic balance can be reduced by moving the hinge line forward, by reducing the balance area, or by adding area to the trailing edge of the rudder. If rudder authority appears inadequate, adding area at the trailing edge will increase both the pedal force required and the authority. Normally a heavier rudder feels better than a light one and results in more accurate control of sideslip.

Creating a good pedal-force gradient will affect other stability and control characteristics and may require other measures to ensure that these characteristics are not degraded. For example, decreasing the rudder's aerodynamic balance to increase pedal forces also increases rudder float, which reduces the effective tail area and degrades Dutch roll characteristics. Conversely, reducing float by increasing the aerodynamic balance also reduces the pedal-force gradient, which is not desirable. A better solution might be to add pedal springs to an aerodynamically balanced rudder to restore the desirable pedal-force gradient while further reducing rudder float.

Although you would expect pedal force to increase with pedal deflection right up to the stops, this does not always happen. Sometimes the pedal force will begin decreasing above a certain rudder angle, as shown in Figure 8.16. In some cases the pedal force can reduce to zero or even reverse. This is called rudder lock and is caused by the stalling of the vertical stabilizer at large angles of sideslip. The simplest cure for rudder lock is to reduce the rudder travel as long as you have sufficient rudder available for takeoffs and crosswind landings. A second good solution is the addition of a dorsal fin, as described in Chapter 7. The dorsal fin reduces the sideslip angle (the real culprit) by increasing the yaw stability, and the vortex coming off the dorsal fin makes the vertical fin more resistant to stall and therefore to rudder lock.

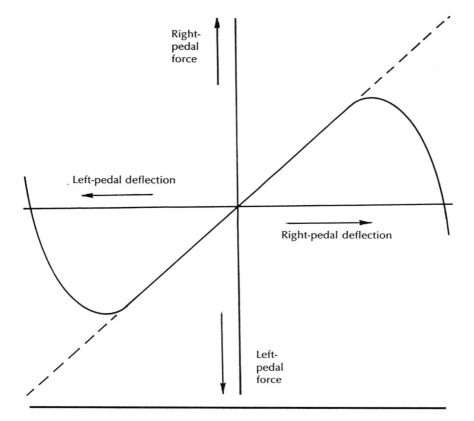

Fig. 8.16. Rudder lock

Summary

As I have suggested, the whole field of stability and control is a complex network of interlocking relationships. Anything you do will affect other things, and a change made to improve one area may very well degrade another.

For this reason it is good practice to complete all your stability and control tests before altering the airplane. Do a series of test points involving at least three speeds (including approach speed and normal cruise) and forward and aft c.g. Then sit down, review the results, and decide what changes to make. Repeat the critical test points and keep making changes until you get the airplane flying the way you want. Then repeat the whole series of points to ensure that you haven't inadvertently created a stability or control problem in an unexpected area. You may find that with clever use of the tools provided in chapters 7 and 8 you can dramatically improve the way your airplane flies with a minimum of effort and expense.

Reference

McClellan, J. Mac. 1989. "Stick to Your Wings Technology: Vortex Generators," *Flying* (October), 22.

Further Reading

Gustafson, David. 1987. "Gordon Price and Unlimited Aircraft's New 10 Dash 200," *Sport Aviation* (March), 12.

Hewes, Don. 1983a. "Effects of Rain and Bugs on Flight Behavior of Tail First Airplanes," *Sport Aviation* (May), 36.

_____. 1983b. "Effect of Rain and Bugs on Flight Behavior of Tail First Airplanes, Part II," *Sport Aviation* (June), 48.

_____. 1983c. "Effect of Rain and Bugs on Flight Behavior of Tail First Airplanes," *Sport Aviation* (July), 61.

Owen, Ben. 1987. "Calling a Spade a Spade," *Sport Aviation* (October), 66.

Sebald, Les. 1984. "Maintenance and Projects," *Soaring* (May), 41.

[CHAPTER 9]

Pitot-Static System

The pitot-static system is one of the simplest in the airplane. Anyone can install a pitot-static system and make it work . . . perhaps. Making it work right is not easy at all, and major manufacturers (especially helicopter manufacturers) spend a lot of time and money calibrating and tuning pitot-static systems. Unfortunately, what they do is as much art as science, and the final adjustments normally consist of cut-and-try on the airplane. A homebuilder does not have to conform to the exacting standards of production airplanes, but you do need to be able to properly calibrate your pitot-static system and make enough adjustments to make it reasonably accurate.

The normal pitot-static system, which is shown in Figure 9.1, measures dynamic pressure (q) and static pressure (p) and delivers them to the cockpit pressure instruments. The dynamic pressure tube should be connected to the port labeled P (for Pitot pressure) on the back of your airspeed indicator. The static pressure tube should be connected to the port labeled S (for static). To prevent confusion, I will use the aerodynamicist's terms q and p in the text. The airspeed indicator measures the difference between q and p, mechanically linearizes the result, and presents it to the pilot as indicated airspeed (IAS). The altimeter uses p to determine pressure altitude, and the rate of climb indicator measures the rate of change of p to determine rate of climb or descent. All this is very simple and would work very well if we were really measuring p and q. Unfortunately, we aren't.

As your airplane flies through the air, it creates a disturbance around it. This disturbance extends in all directions (even forward), is highly complex, and affects p, q, and the direction of flow. This effect alters the measured static and dynamic pressures and leads to errors in all the pressure instruments. Normally these errors are small, but if you set up your pitot-static system incorrectly, they could be quite large. For this reason an accurate pitot-static system actually begins when you are building your airplane.

In selecting locations for the pressure and static pickups, we must pick locations that avoid the pressure disturbance around the airplane as much as possible. This pressure disturbance is at its largest in the area surrounding the wings and propeller. Specifically, the areas above, below, and behind the wings and those in front of and behind the propeller are not good areas in which to locate pitot or static pickups. The disturbance in front of the wing does not extend very far into the relative wind and decreases rapidly near the wingtip. Thus, a very good location for the pitot would be at or near the wingtip and forward of the leading edge a distance of at least half the wing chord. Another excellent location is at the top of the vertical tail or the tip of one tail surface if you have a V tail. This normally places the pitot above the influence of both the wing and propeller. For sailplanes or pushers, an obvious good location from an aerodynamic standpoint is on a probe that positions the pitot forward of the extreme nose.

Besides causing disturbance just by passing through the air, the airplane flies through an angle-of-attack range of about 15 degrees in its normal flight envelope. All dynamic pressure pickups are affected by a change in angle of attack. Thus, if we could suspend our q and p pickups

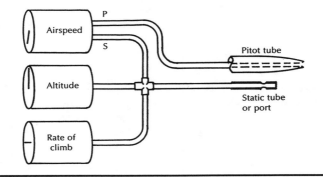

Fig. 9.1. Pitot-static system

well clear of the aircraft interference (on a very long boom, for example), we would still see an error caused by changes in angle of attack. Because the angle of attack varies with the square of the speed, the largest changes in angle of attack, and thus the largest error, occur at the low end of the speed range. Fortunately, some types of pitot systems are less sensitive to angle of attack than others. This factor, and some judicious adjustments, allows us to minimize angle-of-attack errors.

Commercial pitot-static heads are designed to minimize the error due to angle of attack over fairly large angles, so it might be wise to spend the money on a good commercial unit. This has the beneficial side effect that the static ports are built in to the head, eliminating the problem of finding a good static port location. The disadvantage here is increased weight. The angle of attack sensitivity of a simple tube pitot can be reduced by putting a sharp edge on the tube with a drill, as shown in Figure 9.2.

Wing-mounted pitot probes can also be tilted to give a better alignment between the relative wind and the pitot head. Most airplanes cruise at an angle of attack of 2 to 4 degrees and will stall at about 15 degrees unless equipped with leading-edge devices. Simply bending the tube 7 to 9 degrees below the chord line will reduce the low-speed angle-of-attack correction.

This trick doesn't work for tail-mounted pitots because of the complex flow behind the wing. The normal downwash behind the wing will offset the angle of attack and cause the flow over much of the tail to be tilted downward. This will vary with the vertical and lateral position of the pitot tube, and its effect at any given location is difficult to

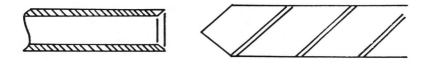

Fig. 9.2. Drilling pitot tube

calculate. The best bet is to locate a tail-mounted pitot as far above the wing as possible and align it with the aircraft's centerline.

Static Port Considerations

If you have chosen a simple pitot tube, you now have to consider the problems surrounding the static port. First of all, venting the static system into the cockpit is a poor idea. Cockpit static pressure almost always differs significantly from outside static pressure, and it often changes wildly with angle of attack. The ideal location for the static port is on a probe clear of the fuselage or wing, as it is on the Aeronca Champion. If you're going to all that work, you might as well install a good commercial pitot-static head. The location of fuselage static ports, should you choose to go that route, is a very tricky business because the pressure differentials being measured are small and the ports are operating well within the influence of the fuselage. Static ports should not be located in areas influenced by the wing or propeller. For this reason, they should not be installed above, below, immediately behind, or in front of the wing or near the prop. Because the static ports are within the flow field of the fuselage, they should not be in areas with strong pressure gradients around the hull. In simple terms, avoid areas of rapid change of fuselage cross-section, such as the nose of a glider or a rapidly tapering rear fuselage. Satisfying these criteria leaves the tail cone, which is normally away from the propeller and wing and has very little taper. Usually two static ports are installed, one on each side of the tail. They are connected, as shown in Figure 9.3, to reduce the effect of sideslip on static pressure.

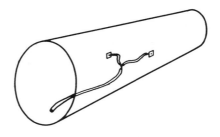

Fig. 9.3. Dual static-port plumbing

Because the static ports are located within the boundary layer, the pressure is strongly influenced by even small local obstructions. For this reason, the static ports should be flush with as few local protuberances as possible, as shown in Figure 9.4.

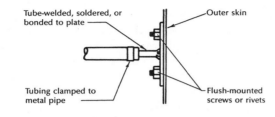

Fig. 9.4. Flush-mounted static port

Airspeed System Calibration

Airspeed systems can be calibrated in one of three ways—a bomb, a calibrated boom, or a speed course. The bomb and the boom are wind-tunnel-calibrated standards that are attached to the test aircraft. The bomb is a finned pitot-static probe (shaped like a skinny bomb), which is hung below and behind the aircraft. It must be reeled out after takeoff and retracted before landing. It significantly restricts maneuverability, so it is normally only used for calibrations.

The airspeed boom is a wind-tunnel-calibrated pitot-static head on a long boom, which is usually mounted on a wingtip or on the nose. An airspeed boom is used for calibration and as an airspeed-measurement standard. As such, it is often left on for the duration of a test program to provide highly accurate pitot-static measurements for test instrumentation.

Neither of these methods is very practical for the homebuilder. This leaves the speed course as the best way to calibrate your system. Unfortunately, the speed course has its limitations, too.

The ideal speed course should be a straight line, on level ground, with no significant surrounding obstructions. It should have two highly visible landmarks, such as crossroads, far enough apart so that it will take at least a minute to fly from one to the other at the highest speed at which you intend to fly. These landmarks will be the entry and exit points of your speed course. The calibration should be done in the lightest winds possible, which means early morning or evening, so it will help if your speed course has no neighbors for you to disturb. A controlled field with almost no traffic, a long runway, and a very cooperative controller would be ideal. Unfortunately, these are rare.

First, measure the distance between your entry and exit points as accurately as you can. Because we are dealing with distances of 1 to 3 miles, our course almost has to be a road. Repeat the measurement several times and average the results to minimize errors. Station a friend on the course with an anemometer and, ideally, a radio. Your friend should record the wind speed and direction for every pass you make.

Fly off to one side of the course and record, with a stopwatch, when the entry and exit points pass under a fixed part of the aircraft, such as the wing's leading edge or a strut. Record the speed and altitude at the entry and exit points, using a table like Table 9.1. If the entry and exit speeds or altitudes differ more than a couple of knots or 20 feet, the data point is no good.

In order to fly this precisely, you must establish your speed and altitude long before you pass over the entry point, and you must move the controls as little as possible while passing between the points. Each speed must be repeated in both directions.

You should fly between 200 and 500 feet AGL. You are going to be concentrating on flying the airplane accurately and on the stopwatch, so you shouldn't be so low that a distraction or momentary loss of altitude could get you too low. It is also good practice to randomize the speed points, that is, to not fly them in consecutive order.

Before discussing how to convert the raw data to useful information, let's review the types of airspeeds we will be working with and the errors built in to each. The parameter just measured with the stopwatch is known as groundspeed (GS), the speed of the aircraft with respect to the earth. If we remove the effect of wind from GS we wind up with true airspeed (TAS), the speed of the aircraft with respect to the air mass through which it is flying. TAS can be converted to indicated airspeed (IAS), the airspeed that we read from the indicator, by applying corrections for density altitude, the pitot-static system, and the instrument itself.

Applying the correction for density to TAS gives an airspeed known as calibrated airspeed (CAS). The difference between CAS and IAS

Table 9.1. Airspeed Calibration

Date _____   Aircraft _____   Pilot _____

Distance _____   Temperature _____   Pressure altitude _____

Run number	Target airspeed (knots)	IAS (knots)	Time (seconds)	Wind speed (knots)	Wind direction	Groundspeed (knots)	Average groundspeed (knots)	CAS (knots)

includes all the corrections relating to the aircraft. These include the error due to the airspeed system (commonly known as position error) and any error from the airspeed indicator itself. It is the difference between CAS and IAS that we are really trying to determine when we do an airspeed calibration.

In case you are really confused at this point, picture the total airspeed calibration like this:

Groundspeed − wind correction = TAS
TAS × density correction = CAS
CAS × system calibration = IAS

After completing your airspeed calibration runs, get together with your friend on the ground and compare notes. Any points that had a wind speed of over 3 knots headwind or 5 knots crosswind should be rejected. For the remaining points, divide the distance between the points by the time to determine groundspeed. Determine TAS by averaging the groundspeed for each pair of runs to eliminate the effect of winds. Correct the resulting TAS for density altitude with your hand computer to yield CAS, using the altitude and temperature you recorded during the speed runs.

The easiest way to examine the results is to plot CAS from the preceding calculations against the IAS you read in the cockpit. If you

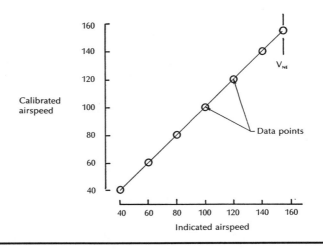

Fig. 9.5. Perfect airspeed calibration

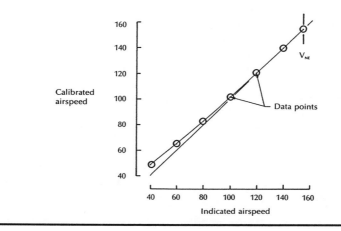

Fig. 9.6. Calibration with low-speed error

have a perfect system, the results will appear as a 45-degree line, as shown in Figure 9.5. Unfortunately, pitot-static systems just aren't this perfect. A more typical calibration is shown in Figure 9.6. This calibration is good at high speed and develops an error of up to 8 knots at the low-speed end. As long as this error is small (3 to 4 knots), this type of calibration is okay. Don't mess with it. In the case of our example, a fix would be needed.

Airspeed System Correction

Errors that require correction normally come in two types—low-speed errors and broad-band errors. Low-speed errors are when the calibration is good at the high end but the error at the low end exceeds 3 to 4 knots. This is normally due to pitot misalignment and can often be improved by increasing the downward tilt on the pitot. Conversely, a good low-speed calibration and a high-speed error indicates a pitot that is bent down too much. This is rare, however.

More difficult to cure is the broad-band error when the error extends through most or all of the calibration, as shown in Figure 9.7. This type of error is normally due to a poor static system.

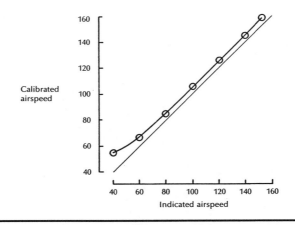

Fig. 9.7. Calibration with high- and low-speed errors

If you have fuselage-mounted static ports, the most obvious fix is to move the ports to a less troublesome location. This is a common approach for production aircraft, but it is a very laborious technique. If you have followed the placement guidelines given earlier, it shouldn't be necessary. As a first pass, however, take a good look at your airplane and see if the static ports are in a position where they could be influenced by the wing, propeller, or local obstructions.

If moving the ports is impractical, another technique can be used to adjust fuselage-mounted static ports. Remember that the static ports must be flush mounted because even small local disturbances can affect the measured static pressure. Local disturbances can work to your advantage, however. By introducing local obstructions called dams, you can adjust the static pressure that the instruments see. Dams are small walls

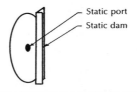

Fig. 9.8. Static dam

about 0.10 inch high located close to the static ports, as shown in Figure 9.8. They can be located forward, aft, above, or below the ports. Normally a dam located in front of the port will increase the indicated airspeed across the whole range. A dam located behind the port will reduce the indicated airspeed. Rotating the dam around the ports will allow you to bias the effect toward the high- or low-speed end of the range. This kind of biasing is very much a cut-and-try game, and it is normally not worth the effort for a homebuilder. The best approach is to temporarily attach the dams with aluminum tape and experiment until you find an acceptable position. Then you should permanently attach the dams to the fuselage.

If you have a pitot-static head or external static probe, the easiest way to adjust it is with a static collar. Find a piece of aluminum tubing that won't quite slip on to the pitot-static probe, and cut from it a sleeve about ¼ to ½ inch long. Cut a slot in the sleeve, clean it up, and slip it onto the probe, gently twisting a screwdriver in the slot to slightly expand the sleeve. When you remove the screwdriver, the sleeve will stay stuck to the probe.

Locating the ring just behind the static ports will increase the static pressure and give a reduced indicated airspeed. Locating the ring in front of the static ports will have the reverse effect. The exact position of the collar is quite critical, and it will probably require several calibrations to get it right. Once you determine the best collar position, bond it in place so that it cannot move.

If you do decide to alter the pitot-static system, make one change at a time and do a complete calibration for each change. Plot the results of the calibrations on one graph in the format used in Figure 9.5 so you can see the effect of each change. Above all, don't get too picky. If you can bring all but the lowest speed points within the 3- to 4-knot error band you are doing very well and should quit.

Further Reading

Bede, Jim. 1972. "Trailing Cone Static Source," *Sport Aviation* (December), 74.
Bingelis, Tony. 1988. "Your Pitot/Static System and Altitude Encoders," *Sport Aviation* (August), 27.
Roemer, Dick. 1986. "Airspeed, What It Is and How to Measure It," *Sport Aviation* (January), 63.

[CHAPTER 10]

Stall-Spin Testing

Stall-spin testing is similar to envelope-expansion flying in that you fly to and beyond a limiting boundary of your airplane. What makes stall-spin testing different is that you will fly close to this same boundary every time you take off or land. This is why stall-spin is still a leading cause of lightplane accidents. The purpose of stall-spin testing is to ensure that your airplane has adequate stall warning and that it can be recovered from a stall or a spin. Stall-spin and flutter testing are the riskiest types of testing you can do, so they both require a careful, methodical approach.

In normal flight, a wing slides smoothly through the air, generating lots of lift and very little drag. As shown in Figure 10.1, the air flows cleanly around the airfoil, and the wing leaves only a narrow wake. When the wing is flying this way, lift is proportional to the angle of attack (the angle between the airfoil and the relative wind) and the square of the speed. Lift is also equal to the airplane's weight times the load factor. In level flight, therefore, the airplane's weight equals a constant times the angle of attack times the square of the speed:

$$\text{Lift} = \text{weight} = \text{constant} \times \text{angle of attack} \times \text{speed}^2$$

As you slow down, V, and therefore V^2, will decrease. If you are maintaining constant altitude, the wing will assume a higher and higher angle of attack, trying to support the airplane's weight. All goes smoothly until the airplane reaches a critical angle of attack. Any further increase, and things change abruptly. The air flowing smoothly over the top of the

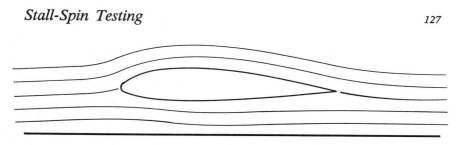

Fig. 10.1. Airfoil in normal flight

wing suddenly separates to form a large, turbulent, drag-producing wake, as shown in Figure 10.2. As this happens, the lift being produced by the wing decreases and the drag increases sharply. The increased drag causes the airplane to slow down, further increasing the angle of attack and making the situation worse. At the same time, the wing develops a strong, nose-down pitching moment. The downwash from the wing, which is proportional to lift, is also dramatically reduced, resulting in a sudden change in the angle of attack of the airflow on the stabilizer. This also pushes the nose down sharply. The sudden deceleration, coupled with the sharp, nose-down motion resulting from the change in wing-pitching moment and the change in airflow over the tail, gives the familiar "falling-out-of-the-sky" feeling we experience in a normal stall.

As you probably know, when a wing stalls, the flow does not break down over the whole wing at one time. The stall usually begins with the flow separating, or detaching, from only a small portion of the wing area, usually at the trailing edge. As the angle of attack is increased, this area of separation grows until it involves most of the upper surface of the

Fig. 10.2. Airfoil in stall

wing. At this point most airplanes will let the pilot know they have stalled by "dropping out of the sky," pitching sharply nose-down, rolling off to one side, or all of the above. The point of initial separation and the growth of the separation determine the characteristics of the stall for your airplane. Fortunately, we have tools that allow us to control the growth of the separated area to get just the stall characteristics we want.

Stalls can occur in two forms, symmetrical and asymmetrical. Symmetrical stalls occur when the same area stalls on both wings at the same time. Symmetrical stalls result in little or no change in roll attitude when the stall occurs, and they probably include all the stall types you were taught as a student. They can be unstable—the airplane tries to recover to prestall conditions—or stable—the airplane remains in the stalled condition so in effect the stall or spin becomes the trim condition. Stable symmetrical stalls, or deep stalls, can be difficult or impossible to recover from.

In contrast to symmetrical stalls, asymmetrical stalls involve a substantial difference between the two wings in either the area or depth of stall, and they are identified by powerful rolling motions. Asymmetrical stalls are either spins or snap rolls, depending on whether they occur in 1-g or accelerated flight.

Let's examine symmetrical stalls first, the stalls you were taught as a stall from level flight or a stall straight ahead. For our stall tests, a slightly different technique is used because the purpose of the test is to determine the prestall and stall characteristics.

Testing for Stall Characteristics

Trim in level flight at some low airspeed, reduce power to idle, and then very slowly reduce the speed, using rudder with little or no aileron to keep the wings level. Be very conscious of the amount of control you must apply to keep the wings level. Carefully observe the speed and any prestall warning the airplane may give you. This may take the form of airframe buffeting, vibration in the controls, or a change in noise. At the first sign of a stall (wing drop or loss of pitch control), recover by dropping the nose and applying full power. If the stall showed good controllable characteristics, repeat the procedure, this time holding the nose up until you observe a clean break before recovering.

Stall characteristics can and do vary from aircraft to aircraft. There are, however, certain characteristics that are considered necessary for safe stall behavior. First, there should be some warning of the stall. Ideally this would take the form of clearly discernible shuddering or buffeting. This occurs when the airflow separates somewhere on the fuselage or wing root and resulting turbulence flows back into one of the tail surfaces. If this does not occur, you should consider installing an electric or pneumatic stall-warning system. Normal practice is to adjust the system so that it activates 3 to 5 knots before the actual stall.

When trimmed for level flight, the airplane should require a noticeable aft stick force to achieve a stall. If you have satisfied the stability and control criteria discussed in chapters 7 and 8, this probably will occur with no further action on your part. Third and most important, the airplane should try to recover by itself. This means that simply releasing the pitch control should result in a reasonable recovery.

There are three classic problems that occur with stalls: wing drop (and yaw), loss of aileron effectiveness, and stable, or deep, stall. The first two are caused by separation beginning at the wrong location on the wing and the last by interference between the wing wake and the horizontal tail.

When a wing stalls, the wing does not suddenly stall over its whole surface. Instead, the stall begins as a separated area at some point on the trailing edge on the wing's upper surface at a speed higher than the perceived stall speed. As the speed decreases, the stalled region spreads until the drag becomes so great that the airplane decelerates rapidly and the stalled area spreads quickly to most of the remaining area of the wing. The point where the stall begins and the way in which it spreads create what is referred to as the stall pattern and largely determine the stall characteristics of the airplane.

Figure 10.3 shows the stall pattern normally associated with good stall characteristics. You will notice that the stall begins at the wing root and spreads gradually forward and outboard as the speed decreases. The aileron is not affected by the stalled region until the airspeed is very low, which allows aileron control to be retained as long as possible. This type of pattern is normally associated with untapered and unswept wings.

Figure 10.4 shows the stall pattern normally associated with a highly tapered wing. The stall begins farther out on the wing and involves the aileron much earlier, resulting in loss of effective aileron control at a higher airspeed. In addition, since the stalled area has a larger moment arm because it is farther from the c.g., small dissimilarities in the stalled area between the left and right wing will cause large rolling moments, normally referred to as wing drop.

If the wing is swept as well as tapered, the stall pattern will look like the one in Figure 10.5. The stall not only begins outboard but spreads outboard more rapidly as well. This leads to even earlier loss of aileron

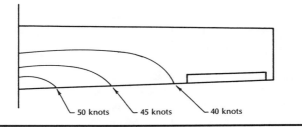

Fig. 10.3. Desirable stall development

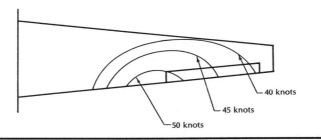

Fig. 10.4. Undesirable stall development—tapered wing

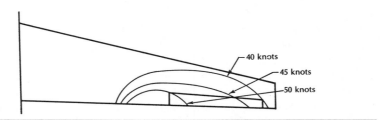

Fig. 10.5. Undesirable stall development—tapered and swept wing

control and a greater risk of wing drop. An additional problem with highly swept wings is the potential for pitch-up when the tip (aft) stalls while the root (forward) is still unstalled.

Tuft Tests

Because wing drop and loss of aileron are caused by the occurrence of initial separation in the wrong place, the most effective tool to determine what is going on is a tuft test. To do a tuft test, tape pieces of wool yarn 6 to 9 inches long to the top surface of the wing. The tufts should be about 12 inches apart both spanwise and chordwise, as shown in Figure 10.6. You will also need a way of photographing the tufts. If you can see them from the cockpit, the most efficient way is to use a camera mounted inside the plane. If the wing is not visible from the cockpit, the photographs will have to be shot from a chase plane.

Take pictures of the tufts at several speeds, including a speed just faster than the first indication of stall and the lowest speed at which you can maintain control. If you have flaps, also take pictures with flaps down.

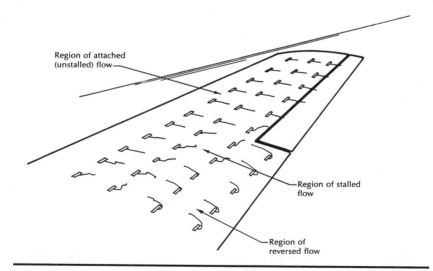

Fig. 10.6. Tufts during stall

As shown in Figure 10.6, the area where the tufts point steadily aft is the area of attached (i.e., unstalled) flow. The stalled region is where the tufts wave around wildly with no discernible direction. You may also see some tufts swing around and point forward on the wing. This indicates a region of reversed flow, which normally occurs deep within the stalled area.

If you have wing drop, first carefully check both wings and ailerons for proper rigging. If the ailerons are rigged slightly down, rerig them to zero or slightly up and repeat the stall. Next, check the surface of the wing in the areas where the stall begins for discontinuities or flaws. On some small airfoils, a spanwise paint line in the wrong place can be enough to cause premature stall. Make an airfoil template and examine the airfoil shape for bumps or hollows in the area of initial stall. If the tuft tests show that the stall is beginning symmetrically but spreading differently on one wing than on the other, fences or vortex generators can sometimes be used to control the growth of the stalled area.

If you find no evidence of local abnormalities that would cause a poor stall pattern, there are several techniques you can use to improve the stall characteristics. The most basic of these is altering the wing twist. Increasing the twist, or washout, by twisting the wingtip's leading edge down tends to move the stall pattern inboard. This makes a tapered-wing stall more like a square-wing. For a cantilever wing or one with a single strut, twisting has to be done in the design phase. However, if you have built a wire-braced biplane or a wing braced with V-struts, it is possible to change the wing twist by changing the wing rigging.

Stall Strips

Another basic technique for improving the stall pattern is adding a device that forces the wing to stall in a more desirable spot at a lower angle of attack. This is done by attaching triangular stall strips to the leading edge of the wing, as shown in Figure 10.7. Once a specific angle of attack is exceeded, the strip trips the flow, causing a stall to begin immediately behind the strip. Adding stall strips is a cut-and-try process. Stick them to your tufted wing with a smooth, strong tape such as aluminum tape. Try your tuft tests and stalls again, observing the effect of the strips. Move or change their length to get the most desirable stall characteristics. Moving them up the leading edge normally causes the stall to occur at a higher speed (and a lower angle of attack), and moving them down

Fig. 10.7. Stall strip

causes a stall at a lower speed (and a higher angle of attack).

Loss of aileron control in a stall is not, in itself, a serious problem. The ailerons should not be used in a stall anyway. This is because if the portion of the wing in front of the aileron is close to a stall, the down-moving aileron will increase the lift coefficient at that point of the wing beyond the critical value, causing a local stall. Because the stall occurs at the aileron (near the tip of the wing), it causes the airplane to roll and yaw sharply, usually away from the intended direction of roll. This is the classic setup for an unintentional spin.

If, however, aileron control is lost at a speed significantly above stall, such as you would see during approach or landing, you have a problem that needs to be fixed. Loss of aileron control at low speeds is usually caused by separation immediately in front of the aileron, as shown in figures 10.3 or 10.4. This can be fixed by adding a cusp or slat at the leading edge in front of the aileron, by increasing the wing twist, or by adding vortex generators. Of the several solutions, vortex generators are the simplest and cheapest. They are also the easiest to add to an existing wing. As for most similar fixes, try a few in the area where you think they will do the most good. Then redo the tuft test to see what you have accomplished.

Stable Stall

Stable stall, or deep stall, is another type of problem entirely. Stable stall occurs when the nose-down pitching moment applied by the elevators and the natural pitch characteristics of the airplane are not sufficient to reduce the wing angle of attack enough to eliminate the stall. Simply put, the airplane's nose does not want to come down enough to break the stall. In fact, the pilot may not have enough control available to force the nose down.

In a normal stall, several phenomena act together to create a strong nose-down pitching moment when the stall occurs. First of all, the types of airfoils we normally use tend to develop a nose-down change in pitching moment when they stall. The second phenomenon comes from the horizontal tail, which, unless it is well above the wing, tends to operate in the wing downwash at low speeds. When the wing stalls, the downwash disappears, resulting in a change in the angle of attack at the tail. This causes a strong upward force on the tail and a nose-down moment. The third component to the nose-down moment is the effect of the elevator as the pilot pushes forward, trying to bring the nose down.

As you can see, there are two things that could mess up this scenario. If the c.g. is far enough aft, the change in pitching moment of the wing may not provide enough moment to push the nose down. In addition, if the horizontal tail is too high, as in a T tail, it may not experience the wing downwash effect at all. In fact, in a stall the wake of the wing sweeps up, as shown in Figure 10.8, often burying a high tail in turbulent low-energy air, making it nearly useless for stall recovery.

You should recognize that the fuselage is unstable by itself, and during all of this it is trying to pitch nose-up. Thus, a successful stall recovery requires a fairly strong nose-down push from the wing or the tail or both. If the c.g. is far aft and the tail becomes buried in the wing wake, there may be no way to supply this nose-down push, and you will find yourself in an unrecoverable stall.

Stable stall is one if the reasons that stall tests are risky and require preparation, altitude, and a parachute. Because c.g. is the only variable you can change easily, stall tests are always begun at forward c.g.

The classic symptom of an impending stable stall is a noticeable reluctance of the nose to drop at stall. If successful recovery requires forcible application of forward stick and/or requires power to help lower the nose, you have a potential stable stall. At the very least it may require restriction of the aft c.g. range. If this is not practical, it may require reconfiguration of the horizontal tail.

A stable or near stable stall is possible with a low-set tail if the tail itself stalls, reducing the elevator effectiveness. The inability to pitch the nose down because of tail stall also normally occurs at excessive aft c.g., because the stalled tail will not develop enough lift to oppose the nose-up moment of the fuselage. The normal solution, of course, is to restrict the aft c.g. range. In addition, strakes can be used to unstall the stabilizer and increase the effectiveness of the elevator.

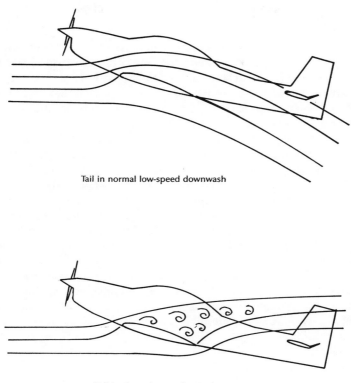

Tail in normal low-speed downwash

Tail in clean air—good nose-down response

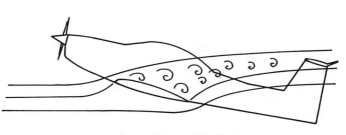

T tail in a stall—loss of tail effectiveness

Fig. 10.8.   Tail positions

Spins

Once you have symmetrical stalls under control it is time to consider the complex world of asymmetrical stalls, i.e. spins. A spin is a condition in which an airplane rotates because one wing is deeper in stall than the other. A spin is a highly complex dynamic maneuver that is still not fully understood, even by the experts.

Before getting into the complexities of spins, it is important to understand what happens to an airfoil after it stalls. Figure 10.9 shows a plot of lift coefficient (C_L) versus angle of attack (α) for a typical general aviation airfoil. The left side of the curve (a) is where C_L varies linearly with α, and is where we normally fly. The point of stall (b) is where a small change in angle of attack causes a dramatic, and sometimes abrupt, change in lift, drag, and pitching moment. Normally our discussion of wing stall ends there, but to understand spins we need to look further. If we continue to increase the angle of attack, the lift coefficient (C_L) continues to decrease and the drag coefficient (C_D) continues to increase. Thus, if speed is held constant and the airfoil is pushed deeper and deeper into stall, lift does not go away but continually decreases while drag is continually increasing. This is where your wing will be operating in a spin.

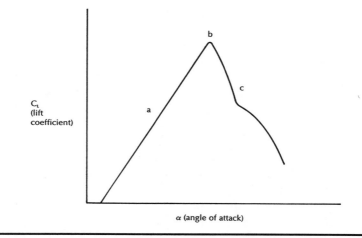

Fig. 10.9. Poststall behavior

In order to grasp some idea of the complexity of a spin, let's examine one in detail to see what happens. As an airplane approaches a stall, the pilot causes the airplane to yaw to the right by applying right rudder. The yaw rate accelerates the left wingtip, reducing its angle of attack. As shown in Figure 10.10, the yaw rate also slows the right wing down, increasing its angle of attack past the critical angle for stall. The flow over the right wingtip stalls, causing a dramatic increase in drag and reduction in lift. This causes the right wing to drop and pull back. The decelerating right wing causes the left wing to accelerate, somewhat like the interaction of ice skaters in crack-the-whip, developing more lift and

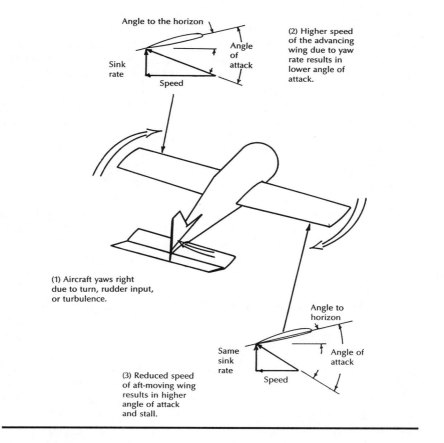

Fig. 10.10. Spin entry

moving the wing farther from stall. At the same time the right wing (now dropping and decelerating) moves even deeper into stall. The result is that the airplane rolls and yaws into the direction of the beginning spin.

As the spin becomes developed, both wings become stalled. The right wing is operating at a much higher angle of attack and thus is developing more drag and less lift than the left wing. Because the airplane is pointing steeply down, this imbalance of forces is translated into a rolling moment, which causes the airplane to rotate violently to the right as it falls. This is called autorotation and is similar to the motion of a maple seed as it falls from a tree.

As you can see, an airplane will spin when there is a substantial difference in the depth of stall between the left and right wings. Put another way, if you make the asymmetrical stall go away, the spin will stop. You can accomplish this by making both wings fly the same, by eliminating the stall entirely or, better yet, by doing both.

The difference in airflow between the two wings is caused by the rotation. The pilot's normal method of controlling yaw rate (rotation) is with pedal. This is why the first step in the normally accepted spin recovery is to aggressively apply full pedal against the spin. Eliminating the stall requires reducing the angle of attack of the airplane, and the normal method of doing that is with longitudinal stick. Thus, step two is to firmly move the stick forward to the trim position. In most aircraft this will result in a rapid recovery into a near vertical dive.

This recovery technique is based on the ability of either or both of the tail surfaces to generate substantial forces during the spin. If either surface is ineffective, the spin would be difficult or impossible to recover from. Normally, loss of control in a spin is caused by the control surface's being in the turbulent, low-energy wake of some other part of the airplane. Because of its position on the airplane, the vertical tail is most likely to be affected. The culprit is sometimes the wake of the wings but more likely the horizontal tail, as shown in Figure 10.11. Historically there are two classic ways of solving this sort of problem: move the vertical tail forward or add vertical fin area. Unfortunately, recent testing has shown that moving the fin forward is not normally effective. Adding fin area does help, however, and the most effective way to add fin area is with a ventral fin. A ventral fin or, better yet, a ventral fin and rudder will be effective at controlling spin problems because the fin will be in clean air below the fuselage. As you remember, the ventral fin is also a good tool for fixing Dutch roll and other yaw stability problems, so you may be able to kill two problems with one fix.

Another situation that can cause loss of rudder control is the so-

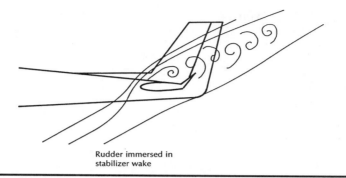

Rudder immersed in
stabilizer wake

Fig. 10.11. Effect of stabilizer on fin in stall

called flat spin. This is a spin in which the nose is higher than you would expect in a normal nose-down spin (45 degrees or less below the horizon). In a flat spin, the fuselage acts like a large low-aspect ratio wing that is stalled. Often the tail is buried in the turbulent low-energy wake of the fuselage and is ineffective. The flat spin is dangerous because the angle of attack is so high that most or all of both wings and the horizontal tail are stalled, making all of the controls useless. You could conceivably wind up with no control authority at all.

Whether or not an airplane spins flat is determined by the aerodynamic and centrifugal pitching moments on the airplane as it spins. Aerodynamic moments are generated by the interaction of the wing and tail with the c.g. As shown in Figure 10.12, the aerodynamic moment generated by the tail opposes the flat spin, while the moment generated by the wing tends to flatten the spin if (you guessed it) the c.g. is too far aft.

Centrifugal pitching moments are a little harder to understand. Although we can treat the mass of the airplane as concentrated at one point (the c.g.) for most performance and maneuvering types of analysis, this doesn't work if the airplane is rotating as it would in a spin. Because each part of the airplane has a mass and a distance from the center of rotation, each part will generate its own forces with respect to the axis of rotation. In the case of a flat spin, one way of understanding this is to envision the mass of the airplane that is forward of the c.g. as concentrated at a point in the nose, and the mass aft of the c.g. as concentrated at a point in the tail, giving a mass distribution somewhat like that of a dumbbell. These two masses generate centrifugal forces, as

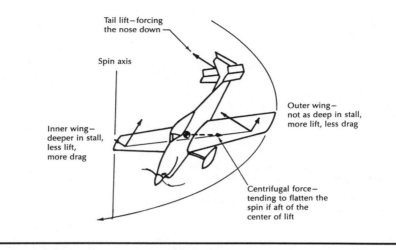

Fig. 10.12. Aerodynamic forces affecting a spin

shown in Figure 10.13. Because the aft (tail) mass is farther from the axis of rotation, it generates a larger outward force, which tends to force the nose up and flatten the spin. The farther these centers of mass are from the c.g., the stronger the flattening tendency. This is why aircraft with large longitudinal mass distributions tend to enter flat spins more easily, while those with the mass concentrated near the c.g. tend to resist flat spins.

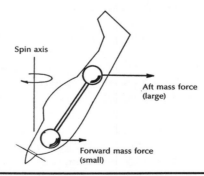

Fig. 10.13. Gyroscopic forces in a spin

Spin Testing Procedures

Now that we have established what a spin is and what can make it get out of hand, let's look at how to test for it. Spin testing is particularly dangerous for two reasons: first, there are no firm, guaranteed-to-work rules. Second, you can't take spin testing a little bit at a time, because with a spin either you are in one or you're not in one. On the other hand, we know that spin characteristics are powerfully influenced by c.g., which does give us a way of minimizing the risk of the initial spins.

To do spins safely, you need altitude, a parachute, a way out of the airplane, and a preestablished and inviolate set of procedures to follow. You will probably be under a lot of mental and physical stress during the spin, and if you have to think too much you will probably do it wrong.

First, set yourself entry and recovery altitudes. For example, you will enter the spin as high as possible (10,000 feet isn't bad), initiate recovery no lower than 5,000 feet AGL, and will bail out at 4,000 feet if the spin has not stopped. This means that at 4,000 feet the rotation must have stopped so that you can begin a normal dive recovery or you get out, right now!

Check that your parachute has a current packing card. Then put it on, as well as everything else you will be wearing when you fly the test. Now go practice getting out of the airplane. Practice by getting in and strapping in as if you were in flight. Then "bail out," using the same procedure each time. For example: pull the mic cord, then pull the canopy release, push the canopy off, pull the harness release, and stand up. Practice it until it's comfortable. If you have an ejectable canopy, get some friends to catch it so you can practice ejecting it without damaging it. If you find any glitches—say you and the chute do not fit through the canopy together, or the parachute pack catches on something in the cockpit—now is the time to fix them. Remember, you are doing this in a quiet hangar with a still aircraft and no wind. In the real world you will be very excited, there will be lots of noise and wind pressure and motion, and your life will be at stake.

If you have not done spins recently, go out with an instructor and practice them until you are comfortable enough so you can observe what is going on around you during the spin. Spins can be highly disorienting if you are not current. In order to cope with potential unknowns, you are going to need to be comfortable enough under very unusual circumstances to remember your procedures and think clearly. Only practice will get you there.

As we have established, the keys to stopping a normal spin are to stop the rotation with rudder and reduce the angle of attack with forward stick. Thus, the standard recovery technique should be: apply full opposite rudder against the spin, wait one half-turn, then shove the stick forward approximately to the trim position until the spin stops. Once the rotation has stopped, neutralize the pedals and recover from the ensuing dive. The sequence of applying the rudder before the forward stick is important. In some airplanes, applying the stick first will accelerate the spin.

Spin testing should begin with the incipient spin. Slow the airplane down, keeping the wings level, and reduce speed gradually until the airplane stalls. As the stall breaks, apply full pedal in the direction of the intended spin. In a conventional spin entry, the airplane will respond by yawing and rolling in the direction of the pedal application. As the wings approach a large angle of bank, recover by briskly applying full opposite rudder and then shoving the stick forward to the trim position. The airplane should stop rolling and start to roll in the opposite direction. The maneuver should end with the airplane's dishing out of the steep bank back to wings-level flight. You should, of course, begin by doing this at forward c.g.

Do the first incipient spins by allowing the airplane to roll to an angle that is comfortable before initiating the recovery. As you gain confidence, use steeper and steeper angles until the wings are at or near the vertical when you apply recovery controls. The application of opposite rudder should stop the roll rate rapidly. You should feel no hesitation and no sensation that the airplane is going to keep rolling into the spin. If this is not the case, stop; then either restrict the airplane or seek a fix before progressing into fully developed spins.

If things are going well at this point, you can let the spin develop to a full turn. Provided no bad characteristics are apparent, gradually build up to three full turns before recovery. There is no point to examining spins beyond three turns unless you are flying a competition aerobatic plane. During this whole process, your airplane should spin nose-down with no tendency for the nose to rise. It should also stop predictably within one turn following the application of full antispin rudder and forward stick. If at any time the nose appears to be rising toward the horizon (indicating a potential flat spin), or the spin takes more than one and half turns to stop rotating (indicating the potential for a stable spin), it's time to stop the test and seek a solution.

If spin characteristics are good at forward c.g., repeat the test at mid c.g. and then aft c.g. If the airplane doesn't want to spin or won't stay in the spin for the desired number of turns, try applying full opposite aileron as you apply spin-entry pedal. This is probably a good thing to try anyway, as it will show the effects of improper aileron usage at low speeds. Sometimes the application of aileron away from the spin will result in a spin that begins and develops much more violently than a normal, pedal-only spin would.

A couple of words of warning at this point. You will note that I have consistently said to move the stick forward to the trim point to effect a spin recovery. In most airplanes with good spin characteristics, simply relieving the back pressure while applying opposite pedal is enough to cause a recovery. The risk to applying full forward stick is that, if the recovery is brisk and you are not quick in getting the stick back again, you risk going inverted and developing substantial negative g's. At the very least this will delay your recovery to level flight.

When the spin stops, you will be in what appears to be a vertical dive, the aircraft will be accelerating rapidly, and there will be a strong tendency to pull back hard to recover from the dive. Take your time. Because the drag of the airplane is so high in a spin, you should not be at a very high speed when you recover. Although the airplane is accelerating, you will have time to do a smooth, even pull-out. The risks of doing a recovery that is too aggressive are overstressing and damaging the airplane or developing a secondary stall-spin. This can be very dangerous, as you will have already used up much of your available recovery altitude.

What do you do if the airplane doesn't respond to the application of spin-recovery controls? First of all, wait. You are going to be all keyed up, and after slamming in full pedal and pushing the stick forward you will expect instant results. It may take several seconds for the rotation to slow and stop. Hold the controls fixed for two full turns before you change anything. If at this point you are still spinning, return the controls to the prospin position (aft stick and rudder into the spin) for one turn; then try opposite pedal and forward stick again. This time get aggressive! Punch in full opposite pedal and full forward stick.

If after applying antispin controls twice you are still spinning, you should rhythmically move the stick forward and aft while maintaining antispin pedal. Imagine that the airplane is falling on a cushion of air. What you are trying to do is get the aircraft rocking on the cushion so that it falls, nose-down, off the edge. Try to feel the frequency at which the airplane is trying to respond and couple with it.

If at this point you are still spinning and still have altitude, apply full aileron into the spin. Some people have recommended to me the application of full power. However, in some airplanes, the application of power accelerates the rate of rotation, so this is not a good idea in an airplane with unknown spin characteristics. In addition, most lightplanes with gravity fuel systems will lose fuel pressure and power during a nose-down spin.

I have just given you more options than you can probably try in an average spin, which is one reason you should plan for lots of altitude. Whatever you do, don't get overly involved in fighting with the airplane. Keep a close watch on your altitude. When you get to your established bailout altitude, don't hesitate—get out! If you know which way the airplane is spinning, bail out away from the spin (if spinning to the right, try to push off to the left). It is highly likely that you will fall faster than the airplane and you don't want it coming down on top of you, so don't wait too long to pull the ripcord. Push free and pull.

If you enter a flat spin, the nose will rise from what appears to be a near-vertical attitude to stabilize much closer to the horizon. You can expect the rate of rotation to change: it may decrease or it may increase as much as 50 to 100 percent. It is likely that one or more of the controls will become ineffective. The best recovery technique you can use is still full pedal against the spin, full forward stick, and, if that doesn't work, use forward and aft stick to rock the airplane nose-down. Again, be careful and watch your altitude, as this can quickly turn into a bailout situation.

Fixing Spin Problems

The easiest solution for a spin problem is to restrict the airplane. This can be done by restricting the aft c.g. limit either all the time, or only for spins, or by placarding the airplane against intentional spins.

Actually fixing a spin problem may be difficult, because the effectiveness of various fixes is not always predictable. The classic way of improving spin recovery is to reduce or control the rotation by adding a ventral fin or fins. Because the fuselage is at a high angle of attack, the ventral fin will be in clean air and will therefore be effective even if the rudder and vertical fin are buried in the wake of the some other part of the airplane. Examples of this approach are shown in Figure 10.14.

Those of you with a tailwheel airplane reap a double benefit because the fin can also serve as a tailwheel fairing.

Strakes can also be used on the horizontal tail, where they serve two purposes. They tend to help pitch the nose down, and they create two vortices that flow up both sides of the vertical tail, energizing the turbulent low-energy flow and increasing the rudder authority.

Another more drastic way of improving both stall and spin characteristics is simply enlarging the horizontal tail. This will have the beneficial side effect of improving pitch stability. It will also require the complete redesign and restressing of the tail structure.

If I were writing this book before 1980, this chapter would have ended here. Until then, what we have discussed up to this point

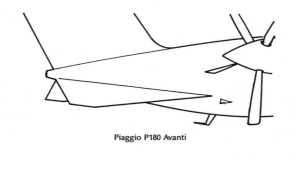

Piaggio P180 Avanti

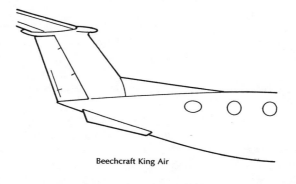

Beechcraft King Air

Fig. 10.14. Ventral fins

constituted the conventional thinking on stall-spin and spin prevention. In 1973 NASA began a program to study spins in light aircraft and found some startling results. The program started with an instrumented Grumman American Trainer equipped with a spin chute, and it was flown at NASA's Wallops Island facility. At one point in the program, the Trainer was equipped with drooped leading-edge cuffs on the outboard half of the wing to improve the airflow over the outer wing and ailerons during the stall. This resulted in a dramatic improvement in the spin characteristics, with the Trainer's becoming almost unspinnable. In order to make the wing modification simple, the inboard end of the cuff had simply been left square. When a fairing was added to the squared-off end to reduce drag, the spin characteristics got worse, almost like the original, unmodified wing. Subsequent testing on the Trainer and three other general aviation aircraft confirmed that the improvement was caused by the working together of the cuff and the step to control the flow over the outboard of the wing.

It is probable that the step in the leading edge forms a vortex that rolls over the top of the wing, preventing the center of the wing from stalling while it prevents the stall on the inboard portion of the wing from progressing outward. The cuff on the outboard portion of the wing in turn prevents the outboard portion of the wing from stalling until it reaches a much higher angle of attack than the inboard portion. Because a normal spin requires that both wings be stalled, the inability to stall a large portion of the wing makes spin entry more difficult, slows the rotation, makes recovery easier and more rapid, and may make a true spin impossible. This has proven to be an even more powerful tool in improving spin characteristics than altering the tail to improve its ability to control yaw rate. Any discontinuity in the leading edge—including steps, notches, teeth, and the so-called vortilons found on the main lifting surfaces of the VariEze (as shown in Figure 10.15)—will probably cause some improvement in stall-spin characteristics, but the most dramatic improvements have been associated with the outboard cuff with a stepped inboard end.

At this point you are probably pretty confused. What started out as a simple problem—stall-spin behavior—has turned out to have more solutions than Noah had animals on the ark. Just how do we go about defining spin behavior and deciding what fixes to use?

The first step, of course, is getting a good wing-stall pattern. If the stall behavior is not acceptable, tuft testing will show what is wrong, and leading-edge stall strips can be used to force the stall to begin where you want it to. If you have a V-strut or wire-braced wing, adding washout can

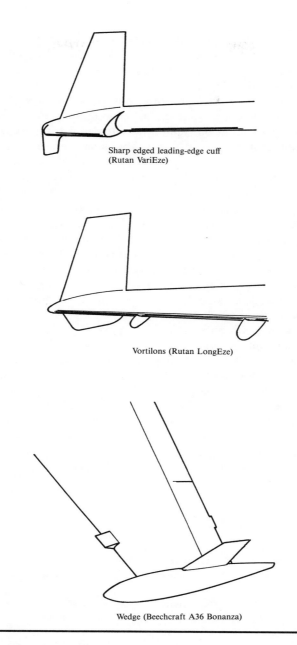

Sharp edged leading-edge cuff
(Rutan VariEze)

Vortilons (Rutan LongEze)

Wedge (Beechcraft A36 Bonanza)

Fig. 10.15. Leading-edge antistall devices

often be used to do the job. If the stall begins in a good place but spreads too fast, vortex generators can be used to keep important parts of the wing, such as ailerons, unstalled, and fences can be used to limit the spread of the stalled region.

Once you are satisfied with the stall, how you address spin characteristics depends on how you want the airplane to spin. For an aerobatic airplane, the spin should be rapid, and entry and recovery should be crisp and predictable. Easy entry and rapid spin require a stall pattern that begins at mid wing and spreads rapidly outboard. This is why aerobatic airplanes usually have swept and/or tapered wings. A crisp, predictable entry and recovery usually requires a powerful rudder, even at high angles of attack. Simply enlarging the rudder may do it, but remember that the vertical fin may be buried in the wake of the fuselage at high angles of attack, reducing the effectiveness of any additional rudder area. A better approach, if you can fit it on the airplane without ground-clearance problems, may be to extend the rudder below the fuselage. This not only increases rudder size in an area that is usually in clean air but also allows room for increased rudder aerodynamic balance if the rudder loads are high.

If you want your airplane to be spinnable but not necessarily aerobatic, then ventral and/or dorsal fins can be used to limit the spin rate and help ensure positive recovery. If there is any tendency toward a flat spin, then strakes can be used on the horizontal tail. You should recognize that all of these improve the basic stability of the airplane, so they can be used to generate improvements in two areas at once if, for example, the yaw stability is marginal. If you want your airplane to have very benign spin characteristics or even to be unspinnable, then you should consider using the drooped leading edge on the outboard half of the wing with a sharp discontinuity where the droop ends.

Further Reading

"All Spins Are Not Equal." 1988. *Soaring* (August), 39.

Anderson, Seth. 1986. "A Critique of the BD-5 Concept," *Sport Aviation* (September), 43.

_____. 1989. "Overview of Stall-Spin," *Sport Aviation* (May), 19.

Bingelis, Tony. 1983. "A Look at Stall Warning Devices," *Sport Aviation* (September), 20.

_____. 1989. "Correction" (to March 1989 article), *Sport Aviation* (April), 10.

Brahney, James. 1988. "New Designs Mean New Spin Rules," *Aerospace Engineering* (October), 15.

Burns, B. R. 1978. "Going for a Spin—Fighter Style," *Flight International* (April), 985.

Chambers, Joseph R., and H. Paul Stough. 1986. "Summary of NASA Stall/Spin Research for General Aviation Configurations," AIAA Paper No. 86-2597 (September).

Cox, Jack. 1988. "New Venture," *Sport Aviation* (November), 13.

DiCarlo, D. J., H. P. Stough, and J. M. Patton. 1980. "Effects of Discontinuous Drooped Wing Leading-Edge Modifications on the Spinning Characteristics of a Low-Wing General Aviation Airplane," AIAA Paper No. 80-1843 (August).

Holmes, Harold. 1984a. "Aerodynamics of Stalls," *Sport Aviation* (April), 10.

———. 1984b. "Exploring Spins," *Sport Aviation* (December), 30.

———. 1985a. "More on Spins," *Sport Aviation* (January), 24.

———. 1985b. "History of Spins," *Sport Aviation* (February), 48.

———. 1985c. "Feedback on Spins," *Sport Aviation* (June), 29.

———. 1985d. "Rudder Discipline in the Pattern," *Sport Aviation* (July), 49.

———. 1985e. "Spin or Spiral Dive," *Sport Aviation* (September), 61.

———. 1985f. "Spinning with the Experts," *Sport Aviation* (November), 61.

———. 1987. "Wing Design and Stall Characteristics," *Sport Aviation* (February), 34.

———. 1989. "NASA's General Aviation Stall/Spin Program," *Sport Aviation* (January), 31.

McClennan, J. Mac. 1989. "Stick to Your Wings Technology: Vortex Generators," *Flying* (October), 22.

Mohler, Stanley. 1987. "They Don't Require Spin Training . . . But Get It Anyway," *Sport Aviation* (September), 40.

North, David. 1984. "Wing Alterations Boost Spin Resistance," *Aviation Week and Space Technology* (July), 143.

Payne, James. 1984. "Spins," *Soaring* (May), 27.

Said, Bob. 1984. "Stalls—The Deeper the Worse," *Soaring* (March), 32.

Schiff, Barry. 1978. "Stall Talk," *AOPA Pilot* (August), 39.

Stough, H. P., F. L. Jordan, D. J. DiCarlo, and K. E. Glover. 1985. "Leading-Edge Design for Improved Spin Resistance of Wings Incorporating Conventional and Advanced Airfoils," NASA Langley Research Center (October).

Twombly, Mark. 1988. "Questair Venture, Two to Go," *AOPA Pilot* (November), 30.

"Vortilons Approved for VariEzes." 1985. *Sport Aviation* (January), 6.

[CHAPTER 11]

Performance

For a production airplane, performance testing is normally the last phase of a flight test program. It requires that the full envelope has been defined and that the aircraft is in a production configuration. The purpose of performance testing is to determine how the airplane should be flown to get the most performance, to satisfy certification or specification requirements, or to substantiate the grandiose claims made by management or the marketing department. Of these reasons, we are really interested only in the first.

Classical performance testing is the process of determining an airplane's so-called power required, or the power necessary to maintain level flight at any combination of speed, weight, and density altitude. This information, when compared with the power available from the engine and propeller combination, as shown in Figure 11.1, allows the calculation of almost all the performance data needed by the pilot. The results of this type of analysis will tell the pilot the performance at any combination of weight, altitude, and speed, and what the optimum performance is and how to achieve it.

For example, as the aircraft goes faster and faster, the power required increases until at point A the power required equals the power available. This is the maximum achievable level-flight speed. Conversely, as you slow down, the power required again converges with the power available. In some lightplanes, especially those with high-drag flap systems, the power available and the power required can cross at a speed above the stall speed (point B), giving a minimum speed below which the airplane cannot be flown without losing altitude. In between these two

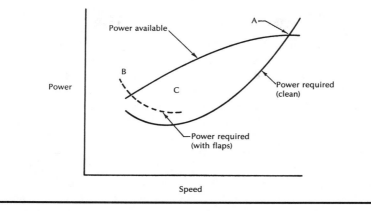

Fig. 11.1. Power required–power available

points, the power available exceeds the power required for level flight. In that range the aircraft can be flown at less than maximum power, or additional power can be used to climb or accelerate. The point at which the difference between power available and power required is the greatest (point C) defines the speed for maximum rate of climb, and the amount of the excess power at that point allows the calculation of the maximum climb rate.

This is a thumbnail description of what hundreds of performance engineers do fifty weeks a year for dozens of aircraft manufacturers worldwide. This kind of analysis is far beyond the capability of most homebuilders and is, in fact, more elaborate than we really need. There are really only three performance parameters that the homebuilder needs to know to fly the airplane safely and get the most performance out of it when necessary. These are best climb performance, best glide performance, and cruise performance. Fortunately, all of these can be adequately determined using simple hand methods.

Climb Performance

By far the most important performance test is for rate of climb. The results of this kind of testing will not only tell you how to achieve maximum climb performance but will also help you set limits, such as the service ceiling and the ability to climb at high density altitude and at heavy weight.

To test for climb performance you are actually going to do a series of timed climbs at fixed airspeeds. All you will need are an altimeter, an airspeed indicator, an outside air temperature indicator, and a stopwatch.

To perform climb tests, first set your altimeter to 29.92 inches of mercury, so you are always dealing in pressure altitude. Start a climb from the lowest practical altitude and climb at a constant indicated airspeed and full power. At each thousand feet, use the stopwatch to record the time to climb from 100 feet below the target altitude to 100 feet above. For example, if the target altitude is 5,000 feet, you should time from 4,900 to 5,100 feet. Use some judgment here; if the rate of climb is very high or very slow, use a larger or smaller increment of altitude to get a reasonable time span. Record the resulting times and altitudes. Be very careful to maintain a constant indicated airspeed and continue the climb to the service ceiling (the altitude at which the rate of climb decreases to 100 feet per minute) or to the highest altitude at which you would conceivably operate the airplane (12,500 feet, for most of us). On the way back down, stop at each target altitude long enough for the outside air temperature to stabilize, and be sure to record it.

There are two things to be careful of during climb tests. Monitor engine temperatures very carefully, especially at low speeds. Engine cooling sometimes gets worse with increasing altitude, and a prolonged climb at low airspeed is the worst possible case. This is where a multipickup head-temperature indicator will pay off. The second thing is the affect the vertical motion of the air mass has on the rate of climb. This may not be detectable from the cockpit. However, we know that a stable air mass exhibits an almost constant temperature lapse rate with altitude. After you land, examine the temperatures you recorded when you were descending. The loss of temperature with altitude should be almost constant (2 degrees C per 1,000 feet is ideal) up to the final altitude. A significant change in temperature or lapse rate indicates the probability of vertical air motion and makes the data useless. It goes without saying that you don't do climb tests when the air is turbulent. As you can see, getting good climb data sometimes requires a lot of patience.

You will need to fly climbs at at least four airspeeds in order to establish the airspeed trend at each altitude. Each speed should be flown twice in order to ensure consistent data. If your airplane can be flown at a range of gross weights, you should repeat the whole process both at maximum gross weight and a minimum normal weight. If you add up all these climbs, you'll see we are now talking about sixteen climbs to the service ceiling, refueling in between, with additional climbs to account for bad data or unstable air. Doing this right will take quite a bit of time. Remember that I said that the FAA minimum number of development hours was probably not enough to do the job right.

As you do the climbs, record the data on a form similar to the one in Table 11.1. Once the flying is done, take your data tables, a calculator, and an aeronautical computer or density altitude chart and start the engineering work. At each target altitude use the outside air temperature and computer or density altitude chart to calculate the density altitude. Next, at each altitude convert the time to climb the 200-foot increment (or whatever you used) into rate of climb as follows:

$$ROC = \frac{\text{altitude change} \times 60}{\text{measured time in seconds}}$$

Plot the rate of climb against density altitude for each speed, as shown in Figure 11.2. Next, for each density altitude, determine the climb rate associated with each speed and plot these, as shown in Figure 11.3. You will note that the high point on each curve gives both the best rate of climb achievable at that weight and altitude and the indicated speed to fly to get that rate of climb. Draw a line through the best rate of climb points (dashed line in Figure 11.3). Then plot the rates of climb and the speeds together, as shown in Figure 11.4. A set of plots or tables giving the information shown in Figure 11.4 should appear in your flight

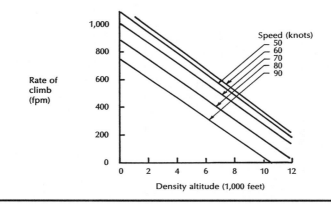

Fig. 11.2. Rate of climb test plot

Table 11.1. Climb Chart

Date _____   Indicated airspeed _____   Gross weight _____

Target altitude (feet)	Altitude change (feet)	Time (seconds)	Cylinder head temperature (degrees)	Outside air temperature (degree C)	Density altitude (feet)	Rate of climb (fpm)

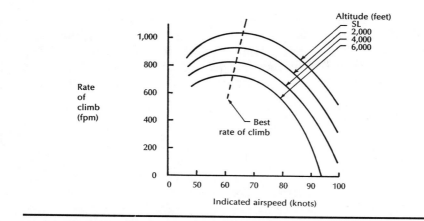

Fig. 11.3. Rate of climb versus airspeed

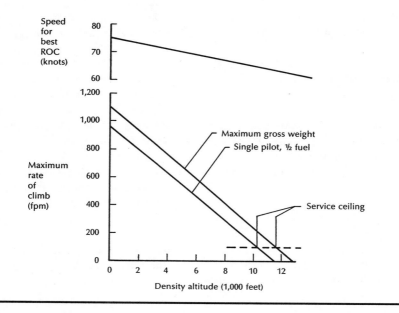

Fig. 11.4. Best climb performance chart

manual. The plot of airspeed for best rate of climb is the most valuable piece of performance data you have. In case of trouble, flying this speed will deliver the best climb performance of which your airplane is capable. Be aware that as your engine gets old and tired the resulting rate of climb will decrease but the airspeed for best climb will remain a constant.

Glide Performance

Glide performance is not quite the same as climb performance because there is no concern for engine power or cooling. The basic technique of timing between two altitudes is the same. Best glide performance is important to give you as much range capability as possible in case of an emergency, such as an engine failure. The important thing to determine is the speed for best angle of glide. If you actually have an emergency, the actual glide gradient will either be good enough or it won't.

It turns out that the glide gradient—often called the glide ratio or the lift to drag ratio (L/D)—and the indicated airspeed for the best angle of glide are nearly constant with altitude. For this reason you can determine glide performance over a range of altitudes by varying the airspeed.

Climb to some reasonable altitude, such as 5,000 feet AGL. Reduce power to idle and start gliding at the minimum airspeed at which you can comfortably control the airplane. Use a stopwatch to determine the time required to glide through a 200-foot increment. Accelerate to a higher airspeed and again time your descent through fixed-altitude increments. Repeat the process for at least four other airspeeds. I would suggest using even altitudes to decrease the chance of errors. For example, if your minimum speed is 55 knots, measure its associated sink rate from 5,000 to 4,800 feet; at 70 knots from 4,500 to 4,300 feet; at 80 knots from 4,000 to 3,800 feet; etc. Repeat the whole process several times, rearranging the order of airspeeds so the sequence is never repeated.

In this case we are looking for the glide gradient, so we will have to do some manipulation of the data to get the results we want. Determine the glide ratio for each point according to the following equation:

$$\frac{\dfrac{\text{Sink distance in feet}}{\text{Time in seconds}}}{\text{Speed in knots} \times 1.689}$$

Plot these points against speed, as shown in Figure 11.5. The lowest point on the curve will give the speed for best angle of glide. This is the number that should be recorded in your flight manual. This is also the speed that you should have locked in your mind as the speed to fly in case of an engine failure.

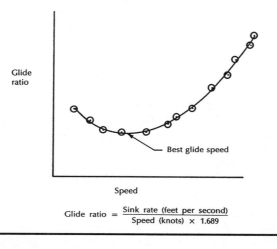

$$\text{Glide ratio} = \frac{\text{Sink rate (feet per second)}}{\text{Speed (knots)} \times 1.689}$$

Fig. 11.5. Determining best glide speed

Cruise Performance

The third type of performance testing important to you as a builder-pilot is cruise performance.

Knowledge of cruise performance helps you accurately estimate how much fuel is required to reach a given destination under normal conditions and how to achieve maximum range if it is required.

The normal production technique of using specialized instrumentation, such as flow meters and strain gauges, is not viable for most homebuilders, so we must fall back upon older methods. For those of you with two-tank fuel systems, this consists of flying carefully controlled cruise segments for a specified period of time on one tank and then, after the flight, measuring the fuel burned from that tank. This process uses a lot of flight time and is therefore expensive. The most efficient way to do cruise testing is to integrate it into your normal cross-country flying after your restrictions are signed off.

Cruise performance can vary with aircraft weight, density altitude, and the power setting used. Working through all of these would be enormously time-consuming. You can reduce the number of variables by taking advantage of the fact that homebuilts are normally only flown over a narrow range of gross weights. If you do cruise testing at maximum gross weight, you will get the most conservative data. An additional factor is that true airspeed and fuel flow vary linearly with density altitude at constant power. For this reason, you should test for cruise performance at at least five power settings (with the highest the maximum normal cruise) and two altitudes 5,000 to 8,000 feet apart. Because of the odd ± 500 rule, you may not be able to pick your altitudes as you would like, but that doesn't matter as long as the spread between them is large enough.

To collect cruise performance data, first carefully fill the fuel tanks and note exactly where the top of the fuel is. Then make sure you are at or near maximum gross weight. If this means ballasting or inviting someone along to increase the weight, so be it. Take off and climb to cruise altitude on one tank, and level off at the desired cruise altitude and rpm. Once established at the cruise altitude, switch to your cruise tank and record your time, rpm, indicated airspeed, altitude, and temperature. If the distance to your destination is short, fly at constant altitude and power until you are ready to descend; if the distance is long, fly more than halfway. Record the time, rpm, speed, altitude, and temperature. Then switch to the original tank before descending to land or continuing the flight. Do not try and run the tank dry! This is not a safe procedure because dry fuel systems sometimes take an appreciable time to refill. Moreover, it may screw up the data because the engine will consume fuel that you cannot account for in the lines and carburetor bowl. After you land, carefully refuel the cruise tank to the same level as before the flight, and note the fuel before filling the other tank.

Record the resulting data on a worksheet, as shown in Table 11.2. Flying a different altitude and speed combination every time you fly cross-country will pile up data fast. For example, flying to and from a fly-in will yield two data points.

Table 11.2. Cruise Performance Table

		Beginning/ending data							Calculations					
Date	Takeoff weight / Fuel	HP	Time	RPM	IAS	OAT	Fuel	Fuel used	Density altitude	CAS / TAS		Average gross weight	Fuel flow

Finish filling in the worksheet by first using your airspeed calibration to convert indicated airspeed (IAS) to calibrated airspeed (CAS). You can use your flight computer, which you probably haven't used since the private-pilot test, to convert pressure altitude to density altitude and CAS to true airspeed (TAS). Now use your pocket calculator to divide the fuel burned from the cruise tank by the time to get fuel flow in gallons per hour.

To convert this data into a usable form, get out some graph paper and sharpen your pencil. First plot fuel flow and TAS against density altitude, using lines of constant power (rpm, if you have a fixed-pitch prop), as shown in Figure 11.6. Remember that fuel flow varies linearly

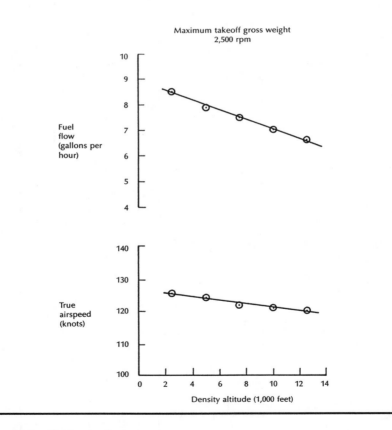

Fig. 11.6. Cruise performance

with altitude, so these lines should be straight. They should also be evenly spaced with power (rpm). If one line is not consistent with the others, you probably have a bad data point and should consider repeating it.

There are three things you want to get from this data: at a given power setting, how fast you will burn fuel, how fast you will go, and how fast you should fly to get the best range. With what you have now, you can generate a set of plots that will tell you all of these.

Set up a sheet of graph paper with speed and fuel flow plotted against power (rpm), as shown in Figure 11.7. From the previous graph, plot lines of constant altitude by going up the speed and fuel flow versus

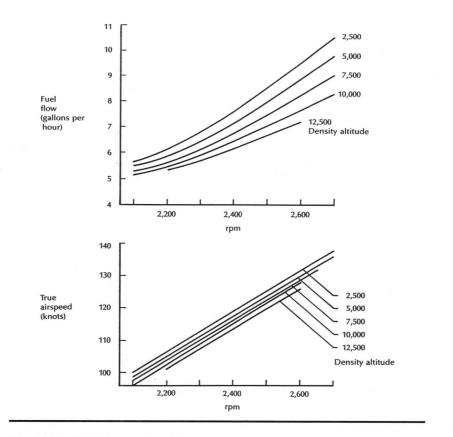

Fig. 11.7. Fuel flow and speed versus rpm

altitude plots at fixed altitudes (2,000 feet, 4,000 feet, etc.). If you normally fly VFR, you can be clever at this point and plot curves for the altitudes you normally fly (2,500 feet, 3,500 feet, 4,500 feet, etc.). These two plots can tell you how fast you will go and how much fuel you will burn at a given altitude and power setting. If, for example, you want to go 200 miles at a cruise altitude of 7,500 feet and 2,500 rpm, you will be burning 8 gallons an hour and going 124 miles an hour. Your flight will take 1.6 hours and you will burn about 10.5 gallons of fuel. If you have 19 gallons of fuel onboard, you will have one hour's worth of fuel for takeoff, climb, approach, and reserve. If, however, you have a 10-mph headwind component, your groundspeed will be 114 miles an hour and your flight will take 1.75 hours and burn 14 gallons. Your 19-gallon tank now only allows you 5 gallons or 0.6 hours for takeoff, climb, approach, and reserve.

This leads logically to the question of how you should fly your airplane to get the required range. If you are not afraid of a little more plotting, you can answer this question. At each altitude and power combination, divide the speed (in miles per hour) by the fuel flow to get a quantity called specific range (in miles per gallon). Plot specific range versus power at various altitudes to get the curves shown in Figure 11.8. Where each curve peaks out on the left side of the plot is the point where this airplane attains maximum specific range. Applying that power (rpm) to Figure 11.7 will give the fuel flow and speed for best range, known as V_{BR}. You will note that this speed is unrealistically low for

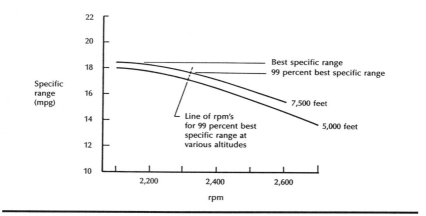

Fig. 11.8. Specific range versus rpm

normal operation, and that the shape of the specific range curve shows that a small sacrifice in specific range gives a rather large increase in speed. For example, if you are willing to sacrifice 5 percent in specific range at 7,500 feet, your TAS will increase 12 percent from 98 to 110 miles an hour. For this reason, aerodynamicists often use a speed known as $.99V_{BR}$. You find this point by taking 99 percent of the peak specific range and finding the corresponding speed. This gives you a much more usable speed with little loss of range. Because the effect of winds is increased at lower speeds, increasing the speed to $.99V_{BR}$ will offset the effect of headwinds, and it often results in the same range as flying at V_{BR}.

As you can see, there are a million ways to play this game. Many homebuilders will see it as too much bother to worry about. If you are willing to put in the effort to do performance testing carefully, though, it will enable you to fly farther and more cheaply in these days of expensive flying. It not only allows you to burn less fuel to get where you are going, but it may also allow you to avoid some time-wasting and fuel-consuming fuel stops.

One word of warning: Using cruise performance data safely requires that you know how much fuel you started out with. Be sure to visually check your fuel before every takeoff.

Further Reading

Bede, Jim. 1973. "Measuring Lift to Drag Ratio," *Sport Aviation* (January), 82.

Bingham, Neil. 1986. "Engine Installation in a Sportplane," *Sport Aviation* (March), 54.

Finch, Reg. 1984. "Wingtip Design," *Sport Aviation* (March), 40.

Gibbons, Robert. 1990. "Performance Modification for Older Wooden Sailplanes," *Soaring* (December), 37.

Henderson, Howard. 1977. "A Study of Cruise Performance of the T-18," *Sport Aviation* (March), 17.

Johnson, Richard. 1989. "Sailplane Performance Flight Test Methods," *Soaring* (May), 26.

Nicks, Oran. 1984. "Drag Awareness," *Soaring* (February), 20.

Schiff, Barry. 1986. "Measuring Performance," *AOPA Pilot* (February), 68.

Sheehan, Eugene. 1984. "Q-200," *Sport Aviation* (March), 21.

Smith, Hubert. 1982. *Performance Flight Testing*. Blue Ridge Summit, PA: TAB Books Inc.

CHAPTER 12

Engine Cooling

High engine temperature resulting from an inadequate engine cooling system is one of the first problems you may encounter in your early flight tests. Engine cooling problems in the first few flights are really serious and need to be fixed before you proceed to further testing. If, however, you get through the initial flights, the problem of high engine temperature is unlikely to rear its ugly head again until you do the prolonged climbs required to determine climb performance. If engine temperature during climbs seems high or if your airplane is equipped with cowl flaps, it is a good idea to do a series of engine cooling tests. They will help you determine your temperature margins and how to fly the airplane to maintain acceptable engine temperatures. You should also remember that engine life is determined, in part, by operating temperature, so the fixes and operating limitations that you apply as a result of temperature tests could reduce your flying costs.

Engine Cooling Testing

The best time to do engine cooling testing is during the climb tests because you are already doing prolonged climbs with maximum power under controlled conditions. Start by adding a column for cylinder head temperature (or oil temperature) to the climb table shown in Table 11.1. When you perform the climb tests, record the head temperature at each target altitude. To process the resulting data, subtract the outside air temperature (OAT) at each altitude from the head temperature and plot the resulting difference against IAS, as shown in Figure 12.1. Full throttle power (and therefore heat produced) for unsupercharged engines and engine cooling at a given indicated airspeed are both proportional to density altitude, so you should get data that varies only with indicated airspeed and configuration (cowl flaps open or closed). You can now use this curve to look at the conditions that are of most interest to you.

The most critical situation for engine cooling occurs during climbout to pattern altitude in a heavy airplane on the hottest days of the year. Under these conditions, combinations of factors gang up on your engine's cooling system. Your engine must produce maximum power (and heat) while you are at best climb speed, which minimizes cooling. Because you are close to the ground, you may not be able to increase speed or reduce power. Because of the heavy weight and high tempera-ture, your climb rate may be poor, so you may have to maintain these conditions for a substantial period of time. The problem is compounded because the highest temperatures occur near the ground.

Let's say your home field elevation is about 1,000 feet MSL and a very hot day is about 33 degrees C, or 20 degrees C hotter than the International Standard Day (ISA) temperature of 13 degrees C. The condition of interest would be at pattern altitude of 2,000 feet and 31 degrees (assuming a normal lapse rate of 2 degrees C per 1,000 feet). Convert the curve of Figure 12.1 to this condition by adding the OAT of 31 degrees to the temperature increase at each speed to give a plot that looks like that in Figure 12.2. Now plot the cylinder head temperature

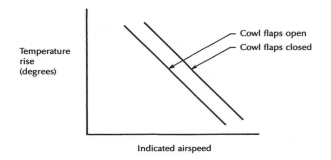

Fig. 12.1. Head temperature increase

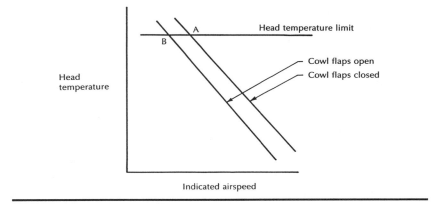

Fig. 12.2. Head temperature at 2,000 feet with ISA + 20 degrees C

limit on this same graph. As you can see, engine cooling is adequate down to the IAS of point A with the cowl flaps closed and down to point B with the cowl flaps open. If the airspeed of point B is close to or higher than the best rate of climb speed you determined in the climb tests, you will not be able to maintain your best climb without over-temping your engine, even with the cowl flaps open. If the best climb speed is higher than point B, this curve will tell you how much cooling margin you have and at what speed you can expect to need to open the cowl flaps.

Most modern engine cooling systems use what is referred to as a pressure cooling system. As shown in Figure 12.3, it has a ram air inlet that pressurizes a compartment usually located above the engine. The pressure differential between this compartment and a low-pressure compartment, usually located beneath the engine, forces the air to go past the engine through a set of baffles. The baffles force the air to pass as close to the hot engine as possible, thus carrying away a maximum amount of heat. The low pressure in the low-pressure compartment is normally created by venting the compartment through an outlet located in a low-pressure region on the outside of the airplane. For a system such as this to work efficiently, there must be a large pressure differential between the two compartments, and the baffling must force most of the air to pass close to the engine.

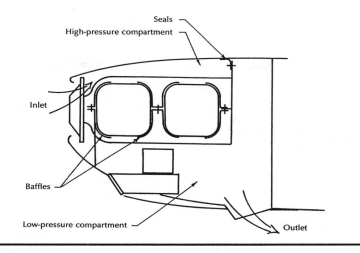

Fig. 12.3. Downflow cooling system

Engine Cooling Fixes

If you do have a cooling problem, first examine the baffles to ensure that they fit closely to the engine and that they don't leak. Carefully inspect the seals around the outside of the baffles. These seals, formed of rubberized fabric and installed as shown in Figure 12.3, seal the baffles to the cowling and prevent air from leaking past the baffles without carrying away engine heat. The seals not only must be in good condition and fit closely to the cowling but also must overlap at the corners to prevent leakage. Note that the seals should be installed curving in toward the high-pressure compartment. A seal installed backwards will be simply pushed out of the way by the high-pressure air. If the seals look good, make sure that the baffling fits closely around the cooling fins on the cylinders.

If there is no sign of leakage around the baffles, the next most likely culprit is not enough difference in pressure between the high- and low-pressure compartments. Normally, the problem is the outlet, which may be too small or may be located in an area where the pressure is not low enough. As a crude rule of thumb, the outlet should be at least 20 percent larger than the inlet in order not to restrict the flow of heated air. If the outlet is large enough, you may need to reduce the pressure at the outlet either with cowl flaps for a high-speed airplane or with a fixed skirt for an airplane with a lower cruise speed.

The skirt can be riveted or screwed to the cowling and then cut down as required to give adequate cooling. Although a skirt is a quick and simple way to improve the airflow through an engine, it has a rather high drag penalty because it creates low pressure by causing separation. In contrast, cowl flaps are complex and require extra attention by the pilot, but they can be closed in cruise, where the heating requirements are smaller, to minimize the drag penalty.

Pusher engines are a chronic source of cooling problems, largely because of the difficulty in finding a good high-pressure location for the inlets. This is because the fuselage blocks the conventional location in front of the engine. Some aircraft, such as the VariEze, solve the ram air problem by extending a scoop below the fuselage. This gives good cooling but has an accompanying drag penalty. Another good location is beneath the wing roots, especially for wings located in the shoulder, or high, positions. This location provides good ram recovery, especially at high angles of attack, by taking advantage of the flow-straightening effect of the underside of the wing and the fact that the underside of the wing is a high-pressure area anyway. In addition, a well-designed inlet can help reduce drag by serving as a partial root fillet.

The added drag of engine cooling can constitute 20 percent or more of the total drag of the airplane. For this reason, the reduction of engine-cooling drag is one area where a clever experimenter who is really interested in performance improvement can show some dramatic results. It turns out that the normal downflow-type of cooling system in a tractor installation has two inherent limitations. Although the normal location of the inlet just behind the propeller is an area of very high pressure, the normal outlet position under the belly is also a high-pressure location. It is also an area in which the pressure increases as the angle of attack increases. This reduces the pressure differential between the inlet and outlet at low speed just when you need the most cooling, as in a climb or slow flight. As described earlier, the normal solution to this is the adding of cowl flaps or a fixed skirt, with the accompanying drag penalty.

The second limitation to the efficiency of the normal downflow cooling system is that it heats the incoming air and then forces it down toward the outlet at the bottom of the airplane. Unfortunately, when air is heated it expands and wants to rise. The pressure differential required to force down the air that wants to rise is greater than if we were able to take advantage of the natural tendency of heated air to go up.

One way of making the engine-cooling system more efficient by countering both of these weaknesses is to use an upflow cooling system. This has been done successfully on the Rutan Quicky and the Defiant and on the T-18 (Garrison 1986). As shown in Figure 12.4, an upflow system takes the ram air into a high-pressure chamber located beneath the engine. The air then passes up into the low-pressure chamber. The outlet can be located above the engine or, as in the case of a low-wing airplane, out the sides of the fuselage above the wing. In cruise, the top of the cowling is not an area of very low pressure. However, as the airplane slows down and goes to a higher angle of attack the pressure above the cowling normally decreases, giving a higher pressure differential and improved cooling just when it is needed. In addition, when the engine is shut down on the ground, the hot air from the engine now has a route to escape, improving cooling on the ground and reducing the chance of vapor lock for fuel-injected engines.

The one problem resulting from an updraft cooling system in a tractor installation is that any contamination (oil, for example) in the cooling air winds up on the windshield. This will probably require more frequent cleaning of the windshield, and that leaking prop seal will need

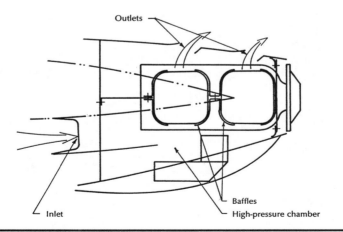

Fig. 12.4. Upflow cooling system on a pusher engine

to be fixed immediately.

An updraft cooling system works even better on a pusher engine. In this case, the top of the cowling is a low-pressure area and the underside of the wing a high-pressure area, giving a maximum pressure differential and increasing pressure as the airspeed decreases (and angle of attack increases). It will probably eliminate the need for large, draggy cooling scoops, and the windshield contamination problem won't exist.

Reference

Garrison, Peter. 1986. "A New Cowling for the T-18," *Sport Aviation* (September), 44.

Further Reading

Bingelis, Tony. 1974. "Cowling Installation for the Homebuilt," *Sport Aviation* (February), 51.

Bingham, Neil. 1986. "Engine Installation in a Sportplane," *Sport Aviation* (March), 54.

Blackstrom, Al. 1974. "Toward Better Performance," *Sport Aviation* (November), 31.

Bond, Dan. 1989. "Introduction to Cooling System Design," *Sport Aviation* (August), 39.

Owen, Ben. 1989. "Engine Cooling Tips," *Sport Aviation* (August), 59.

Taylor, Molt. 1988. "Cool It," *Sport Aviation* (April), 40.

Thurston, David. 1978. "Powerplant Installation," *Sport Aviation* (May), 56.

[CHAPTER 13]

Conclusion

You now have all the basic tools required to do an adequate flight test program for a lightplane. The one remaining question is how to organize all of this into a coherent plan. In doing so you should observe three important criteria.

The first criterion, of course, is safety. This means the early and methodical development of a usable envelope and a careful buildup to any high-risk maneuver. This also means not taking too big a bite out of the envelope on a single flight.

The second criterion is to make efficient use of your flight time. Flying time is expensive, and boring holes in the sky may be fun but it wastes money. Recognize that doing a test program properly will probably use more time than the FAA requires, and this will cost you. Write each flight card so that you can do tests during as much of your time in the air as possible. For example, use your climb up to test altitude to accumulate climb performance data or to record engine cooling data, or both. Upon reaching the test altitude, first do any tests required to check the effect of changes to the airplane since the last flight. Follow this with twenty minutes of tests, such as those for handling qualities, that can be done within the established envelope. Finish the flight with the high-risk points, such as flutter tests, or those that will require an inspection after landing, such as envelope expansion. On the way back down, do stalls or glide performance tests.

The third criterion is to do whatever may require changes to the airplane as early in the test program as possible. The best example of this is handling-qualities testing, which may result in the addition of strakes, dorsal or ventral fins, or modifications to the control system.

If you consider all of these, you wind up with a test program that looks something like this:

1. Develop the load factor envelope up to maneuvering speed.

2. Do stability and control tests within the established envelope while enlarging the envelope a little on each flight. Add flutter tests to the envelope expansion once you are above 140 knots. Continue to V_{NE}.

3. Calibrate the airspeed system.

4. Complete the stall tests.

5. Using the airspeed calibration, test to 110 percent V_{NE}.

6. Do spin tests.

7. Do performance testing incorporating engine cooling in the climb tests, if needed.

As you can see, this sequence roughly follows the chapters of this book. Each group of tests prepares the way for the one following. The culmination, the performance tests, shows you the true potential of the airplane and how to achieve it. You may modify this sequence to fit your particular circumstance, but be sure you have a well-thought-out reason for doing so.

We come full circle with a word of warning. Cautious, methodical development equates to safety. Doing aerobatics on the first flight or pushing out to V_{NE} on the second may be macho, but it's not smart. "The right stuff" has killed more test pilots than any other single factor. Carefully think out what you are going to do ahead of time, on the ground. If something unexpected happens or the airplane doesn't feel right, discontinue the test, come back and land, and check it out.

By the same token, flight testing will never be safe in the sense the lawyers and politicians use. There will always be a significant risk. It's up to us, as pilots, to drive that risk down as low as possible.

When the test program is done, you will understand your airplane better than anyone on the face of the earth. You will know how far you can push it and how to fly it to get the best performance from it. You will also have developed a level of piloting precision, aeronautical knowledge, and cool, balanced judgment you never expected to attain. This is the mark of the test pilot.

Further Reading

Bingham, Neil. 1988. "Design Analysis, A Critical Analysis of the KR-2," *Sport Aviation* (January), 38.

Brechner, Berl. 1976. "Test Pilot," *AOPA Pilot* (August), 32.

FAA. 1990. *Amateur-Built Aircraft Flight Testing Handbook.* Washington, D.C.: U.S. Government Printing Office.

Schiff, Barry. 1976. "The Road to Certification," *AOPA Pilot* (August), 28.

[INDEX]

The purpose of this book is to give you, the builder, the tools to test your creation safely with professional results. . . . If you can discipline yourself to flight test properly, you will increase your knowledge of your airplane and its capability tremendously. — Introduction

Flight Testing Homebuilt Aircraft

VAUGHAN ASKUE

The majority of experimental aircraft flown each year, says author Vaughan Askue, are homebuilt planes tested and piloted by their builders. *Flight Testing Homebuilt Aircraft* offers the homebuilder safe, effective techniques for flight testing, techniques that are often only touched on in journals or derived with difficulty from technical manuals. Taking a step-by-step approach, Askue leads the homebuilder through flight preparation, taxi tests, first flight, and envelope expansion, discussing common stability and control problems and ways to resolve them. A compendium of practical tips and oral lore on reducing risk and improving performance, this book is a "must-have" resource for pilots and builders.

VAUGHAN ASKUE, manager of customer development at Sikorsky Aircraft in Stratford, Connecticut, is a longtime pilot and a former flight test engineer. He holds a B.S. in aeronautical engineering from Rensselaer Polytechnic Institute. An active member of the Experimental Aircraft Association and the Sailplane Homebuilders Association, Askue has worked on the design and construction of several homebuilts.

For a complete list of
our aviation titles, contact
IOWA STATE UNIVERSITY PRESS
2121 S. State Ave.
Ames, Iowa 50010-8300
Telephone 515/292-0140
FAX 515/292-3348.

ISBN 0-8138-1308-5

90000>

9 780813 813080